T0233843

The Smart Grid

Adapting the Power System to New Challenges

Synthesis Lectures on Power Electronics

Editor

Jerry Hudgins, *University of Nebraska, Lincoln*

Synthesis Lectures on Power Electronics will publish 50- to 100-page publications on topics related to power electronics, ancillary components, packaging and integration, electric machines and their drive systems, as well as related subjects such as EMI and power quality. Each lecture develops a particular topic with the requisite introductory material and progresses to more advanced subject matter such that a comprehensive body of knowledge is encompassed. Simulation and modeling techniques and examples are included where applicable. The authors selected to write the lectures are leading experts on each subject who have extensive backgrounds in the theory, design, and implementation of power electronics, and electric machines and drives. The series is designed to meet the demands of modern engineers, technologists, and engineering managers who face the increased electrification and proliferation of power processing systems into all aspects of electrical engineering applications and must learn to design, incorporate, or maintain these systems.

The Smart Grid: Adapting the Power System to New Challenges
Math H.J. Bollen
2011

Digital Control in Power Electronics
Simone Buso and Paolo Mattavelli
2006

Power Electronics for Modern Wind Turbines
Frede Blaabjerg and Zhe Chen
2006

© Springer Nature Switzerland AG 2022

Reprint of original edition © Morgan & Claypool 2011

All rights reserved. No part of this publication may be reproduced, stored in a retrieval system, or transmitted in any form or by any means—electronic, mechanical, photocopy, recording, or any other except for brief quotations in printed reviews, without the prior permission of the publisher.

The Smart Grid: Adapting the Power System to New Challenges

Math H.J. Bollen

ISBN: 978-3-031-01368-3   paperback
ISBN: 978-3-031-02496-2   ebook

DOI 10.1007/978-3-031-02496-2

A Publication in the Springer series
SYNTHESIS LECTURES ON POWER ELECTRONICS

Lecture #3
Series Editor: Jerry Hudgins, *University of Nebraska, Lincoln*
Series ISSN
Synthesis Lectures on Power Electronics
Print 1931-9525   Electronic 1931-9533

The Smart Grid
Adapting the Power System to New Challenges

Math H.J. Bollen
Luleå University of Technology, Skellefteå, Sweden,
STRI AB, Gothenburg, Sweden,
Energy Markets Inspectorate, Eskilstuna, Sweden

SYNTHESIS LECTURES ON POWER ELECTRONICS #3

ABSTRACT

This book links the challenges to which the electricity network is exposed with the range of new technology, methodologies and market mechanisms known under the name "smart grid". The main challenges will be described by the way in which they impact the electricity network: the introduction of renewable electricity production, energy efficiency, the introduction and further opening of the electricity market, increasing demands for reliability and voltage quality, and the growing need for more transport capacity in the grid.

Three fundamentally different types of solutions are distinguished in this book: solutions only involving the electricity network (like HVDC and active distribution networks), solutions including the network users but under the control of the network operator (like requirements on production units and curtailment), and fully market-driven solutions (like demand response). An overview is given of the various solutions to the challenges that are possible with new technology; this includes some that are actively discussed elsewhere and others that are somewhat forgotten.

Linking the different solutions with the needs of the electricity network, in the light of the various challenges, is a recurring theme in this book.

KEYWORDS

smart grid, power transmission and distribution, renewable electricity production, electricity market

For Irene

I want the future now
I want to hold it in my hands
I want a reason to be proud
I want to see the light

Peter Hammill, *The future now*, 1978

Contents

Preface and Acknowledgments

This book was written during a two-month period (July and August 2011) and is about the smart grid, from the viewpoint of the electricity network ("the grid"). Many discussions within the field of power systems (and even some far beyond that field) are about the smart grid, especially about what it is and what it is not. In this book, I present the smart grid as "what it should do". For that purpose, I have translated some of the developments in society into challenges for the grid. On the other side, new technologies and methodologies are translated into solutions to address these challenges. The most important challenges are the ones due to the required transition to a sustainable energy system. There is significant emphasis on these in the book.

The technical details of the different new technologies and methodologies are not discussed in this book, neither are the developments in society discussed in detail. In this book, I have really tried to concentrate on the "grid" in "smart grid", with a result that the "smartness" may appear somewhat left out. But I have no doubt that there will be lots of books and even many more papers and technical reports addressing those technical details (there already are, in fact).

The material and ideas presented in this book originate largely from the work I was allowed to do for the Energy Markets Inspectorate, for STRI, and for Luleå University of Technology. These three organizations deserve much credit for this. The content, especially the inaccuracies and controversial opinions, remains completely my own responsibility.

A word of thanks goes to the many people that inspired me while working on various aspects of the smart grid, ranging from colleagues close and far, to participants in different projects, presenters at various conferences, and authors of all kinds of publications.

My special thanks go to Nicholas Etherden, Sarah Rönnberg, and Aurora Gill de Castro for their critical review of the manuscript.

Math H.J. Bollen
September 2011

CHAPTER 1

Introduction

The electricity grid finds itself at the start of what might become big changes. These changes are driven by a number of developments in society, where the transition to a sustainable society is the most important one. Without serious changes in the way in which the grid is designed and operated, it may become a barrier against this transition. The changes in society translate in new challenges for the electricity network. Using existing technology will not always allow these challenges to be solved, and even when it is possible to solve them, new technology may do the job better and cheaper.

In this chapter, Section 1.3, we introduce those new challenges, and we introduce the solutions based on new solutions in Section 1.5. Those solutions are often referred to as the "smart grid", a term that is explained further in Section 1.2. But first we will introduce the electricity network in Section 1.1 and the different stakeholders in Section 1.3.

1.1 THE ELECTRICITY NETWORK

The aim of the electricity network is to connect consumers and producers of electrical energy. The term "network user" refers to both consumers and producers and to those that sometimes produce and sometimes consume electrical energy. The latter are also called "prosumers".

In economic terms: "The grid should be an enabler of the electricity market". In technical terms, the primary aims of the power system are to ensure acceptable reliability and quality for all network users without any intervention in the power flows that result from the electricity market. Next to the primary aims, a lot of secondary aims are a result of the way in which the grid is designed and operated. Examples of secondary aims are the following: preventing overloads; maintaining sufficient operational reserves and preventing incorrect operation of the protection. The secondary aims are important for keeping the primary aims, but the secondary aims do not directly impact the network users. For example, to maintain a high reliability, the network operator maintains reserves. But if the same reliability can be obtained with less reserves, there is no longer any need to keep those reserves. This is an example that will return several times in this book.

The terms "grid" or "power grid" and "system" or "power system" are sometimes used as a synonym for electricity network, sometimes in a slightly different meaning. The term "power system" refers to the combination of the electricity network and all electrical equipment with the network users. The power system covers all the electrical interactions between equipment and the network, from the largest generator to the smallest motor and everything in between. Whereas the "network" would only consist of the wires, the "system" also includes the equipment that is connected to the wires. The term "grid" is used as a synonym both for electricity network and for power system. In this book, we will use the three terms somewhat as a synonym where the difference in meaning will be emphasized when needed.

1.2 THE SMART GRID

The term "smart grid" has appeared often within the last few years in technical as well as in nontechnical literature. A search after the term "smart grid" in a number of databases with technical publications resulted in Figure 1.1. Before 2004, the term is only rarely used, but from 2005 the use of the term grows very fast. Using Google's search engine gives between 50 and 100 thousand hits per year up to 2005, 2 million hits for 2009 and almost 9 million hits for 2010.

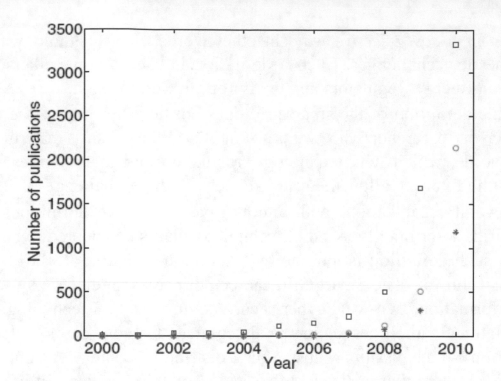

Figure 1.1: Number of publications on "smart grids" according to Scopus (star), IEEE Xplore (circle) and Google Scholar (square)

Next to "smart grids", alternative terms are also being used like "intelligent network", "the energy internet" and "wise wires", but the term "smart grid" is by far dominating in its use. The term is used to refer in one way or to other to the future electricity network. Often reference is thereby also made to new challenges that this smart grid is able to cope with much better than the existing grid. The reason that the smart grid is better in this is that it uses the newest technology, for example, communication and power-electronics control, but also because of much more active customer involvement in the operation of the grid than in the existing situation. In Section 1.4, we will introduce these new challenges, and in Section 1.5, an overview is given of the new technology. Further details of the challenges are given in Chapter 2 and of the new technology in Chapter 3 through Chapter 5.

Many definitions and descriptions are given of what the smart grid is. The definition used in this book is completely technology neutral: *smart grids are the set of technology, regulation and market rules that are required to address the challenges to which the electricity network is exposed in a cost-effective way*. In other words: new technology to cope with the new challenges. Such a definition is used among others by the European Energy Regulators and the European Commission.

This definition of the smart grid does not however satisfy the curious engineer or researcher who wants to know, "What is this smart grid?" In other words, which new technology, regulation and market rules will be used in the future to allow the grid to cope with those challenges?

Several organizations and authors give alternative definitions of the smart grid, describing the kind of technology that is considered to be part of it. One such definition is the one by Gharavi and Ghafurian [2011] which reads as follows: "The Smart Grid can be defined as an electric system that uses information, two-way, cyber-secure communication technologies, and computational intelligence in an integrated fashion across electricity generation, transmission, substations, distribution and consumption to achieve a system that is clean, safe, secure, reliable, resilient, efficient, and sustainable".

Before we continue, a few things should be set straight. The first thing is that the smart grid does not exist yet; neither does anybody know what the smart grid will look like. It is also important to realize that we do not talk about replacing the existing electricity network with a new one. Parts of the network will be replaced with something completely different, others will be replaced with something similar and some parts will not be replaced at all. Finally, there is no sharp transition between a "non-smart grid" and a "smart grid". Some of the developments to be discussed later in this book already started 50 years ago whereas others might take another 10 years before they are ready for practical implementation.

The reason for giving a technology-neutral definition is that we do not want to rule out any technology beforehand. The choice of technology depends on the specific challenge and on many local circumstances. Something that is "smart" at one location does not have to be smart at another place.

In the remainder of this book, we will not so much refer to the smart grid but instead describe the different methods and technology that might become part of the smart grid: in other words, the new technology for the new challenges.

1.3 THE STAKEHOLDERS

The electricity network connects producers and consumers of electrical energy. A distinction is often made between different voltage levels in the electricity network. The transmission network, with the highest voltage level, a few hundred kV, covers the whole country and connects the largest power stations with the main centers of consumption. The distribution network, at the lowest voltage levels, up to about 30kv, delivers the electricity to the smallest consumers. An intermediate level, the subtransmission network, is distinguished at voltage levels around 100kv. Most production units are connected to the transmission and subtransmission network and most consumption is connected to the distribution network. Some large industrial consumers are connected to a higher voltage level, some smaller production units are connected to a lower voltage level.

In some countries, the network is owned by one company or by a small number of companies; in other countries, the electricity network is a patchwork of hundreds of small and large companies. No matter their size, these companies are called "network operators" or "utilities". A utility owns not only part of the network but also all or most of the production units connected to it. In many countries (including almost all European countries), this is no longer allowed: the network operator should concentrate on the business of owning and operating the network and not discriminate between producers. Even when the network operator does not own production, a system operator is needed to take care of the operation of the network together with the production units. A "transmission-system operator" also owns the transmission network, whereas an "independent system operator" does not own any assets.

The system, transmission and distribution network operators are also referred to as the regulated part of the electricity market. They are natural monopolies; a customer cannot choose between network operators (unless moving to another town) so there is no open market for network operation. A network operator has the right as well as the obligation to connect producers of consumers within its service territory. To prevent that the network operator will misuse its monopoly, the connection and use-of-system tariffs are under the control of a regulator.

Whereas the transport of electricity takes place on a regulated market, its trading takes place on an open or deregulated market. This is shown schematically in Figure 1.2. The electricity market consists of a wholesale market and a retail market. At the wholesale market, the producers sell their electricity and this is bought mainly by electricity retailers. Also some large consumers buy their electricity directly at the wholesale market. The wholesale market typically consists of a day-ahead market where trading is based on predicted consumption and production and a real-time or balancing market where an adjustment to the price is made based on the actual production and consumption. The workings of the wholesale market are explained in more detail in Chapter 5.

The retailers buy electricity at the wholesale market and sell it to the consumers at the retail market. Also some smaller producers sell their production at the retail market. At the wholesale market, individual players place selling and buying bids after which a market price is obtained through a set of pre-defined rules. At the retail market, there are bilateral deals between one retailer and one consumer who agree on a price for a longer period of time, up to a few years. Competition is ensured by allowing consumers to freely choose between retailers and to change retailers as well.

An important difference between the wholesale market and the retail market is that the price at the wholesale market varies strongly with time, whereas these price variations are invisible for the consumers on the retail market. Exposing individual consumers to the price variations on the wholesale market is one of the smart-grid solutions to be discussed in this book.

The various developments that are discussed in this book will result in some more stakeholders getting into the picture. These new stakeholders include the new types of production units, mainly based on renewable energy but also small and medium-sized combined-heat-and-power (co-generation) units. These units behave in a different way than the existing production units, which are in this context referred to as "conventional production units". Production units not only inject electric power into the grid, they also support the grid in a number of other ways, through so-called "ancillary services", for example, frequency control. The introduction of new types of production makes it no longer obvious that these ancillary services are also provided by the production units. Instead, an "ancillary-service provider" could enter the market. In most countries, markets already exist for frequency control, but other ancillary services are often compulsory as part of the connection agreement for a production unit. Another of the solutions to be discussed later is creating markets for more of the ancillary services, as it is no longer obvious that it is most cost-effective to let them be provided by the production units.

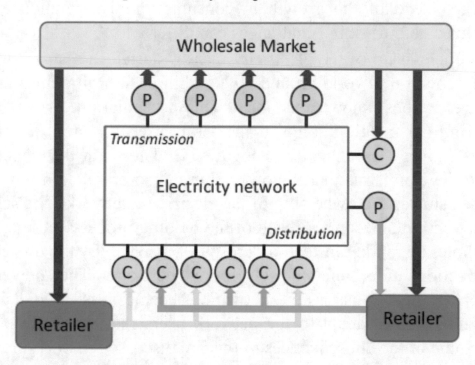

Figure 1.2: The (regulated) electricity network and the (deregulated) electricity market. P = production unit; C = consumer.

It is expected that in the future, small network users (consumers, producers and those that do both) will get more involved in different markets, from the day-ahead market through ancillary-service markets. Getting involved directly in such markets will only be possible for the biggest network users (large industrial consumers, large wind farms). Smaller customers will most likely become involved through an "aggregator". Such an aggregator will play a similar role as the retailer for the wholesale market; in fact, it may well become the existing electricity retailer that takes on many of the aggregator functions.

1.4 THE CHALLENGES

The challenges to which the electricity network is exposed vary between different countries. In most industrialized countries, the main challenges are related with the introduction of an open electricity market and with the transition to a sustainable energy system. In developing countries, the main challenge is more likely to be how to get electricity to the all consumers, whereas in countries with a fast growing economy, the main challenge may be to keep ahead of the growth in consumption. Even among different industrialized countries, the challenges can differ a lot.

The opening up of the electricity market started in a small number of countries around the world around 1990: Chile, The Philippines and Great Britain are the best-known examples. This deregulation, privatization, or restructuring (to mention some of the terms used) created the electricity market as shown schematically in Figure 1.2. The main difference from a technical viewpoint was that the ownership of the production units was no longer the same as the ownership of the electricity network. The scheduling of production units became a result of the bidding process and market rules on the wholesale market instead of a result of dispatch by the owner of both the production units and the electricity network. The power flows, especially in the transmission network, became less predictable. The network now had to adapt to the network users instead of the other way around. Some of the consequences of this are discussed in Section 2.3.

The transition to a sustainable energy system will require more efficient production and use of energy sources and also an increased use of renewable sources of energy. Both will have their impact on the electricity network. New types of production units will appear based on renewable sources but also combined-heat-and-power units. These units are often smaller than the conventional units and are therefore connected to parts of the network where earlier only consumption was connected. But even when it concerns large units connected to the transmission network, they behave in a different way than the conventional units. We will discuss this in detail in Section 2.1. Energy efficiency can result in less consumption of electricity but also in more consumption of electricity. Less consumption could occasionally be a challenge as well, and some types of more efficient equipment could have a negative impact on the voltage quality but the main challenge will be the introduction of new types of electrical equipment replacing less efficient non-electrical equipment. The electric car is often mentioned as an example, but a shift from gas to electrical heating and cooking would actually be a much bigger challenge to the grid and could take place much faster. This will be the subject of Section 2.2.

Two of the old challenges facing the electricity network, maintaining reliability and providing sufficient transport capacity, are important in many developing countries and in countries with fast-growing economies. But also some of the earlier challenges have an impact on this; for example, large amounts of renewable electricity production will require more transport capacity, and they could also endanger voltage quality and continuity of supply. More on this in Section 2.4 and Section 2.5.

The electricity network has adapted itself to new challenges throughout the 100 years of its existence. Whereas it started as a small low-voltage dc network, it developed into a very large ac network with many voltage levels so as to increase reliability and get access to many more sources of energy. Hydropower at remote locations resulted in the development of even higher voltage levels (400kv in Sweden and 750kv in Canada) and HVDC. Building large nuclear power stations required the solving of many stability issues. Higher demands on reliability led to operating routines with primary, secondary and tertiary reserves.

A very important challenge, and maybe the most difficult one and the one most different from earlier challenges, is that the future energy mix will be very difficult to predict. Solar power features high in many long-term scenarios, and there remains the promise of nuclear fusion. The number of supporters for nuclear has reduced a lot recently; this again confirms the unpredictability of the future mix. See, for example, Kaku [2011] for a vision on how our energy supply will look like in 2100, and, among many others, Abbott [2010], Johnson [2009], Friedmann [2009], Jacobson and Delucchi [2009], and the excellent book by MacKay [2009] for a view at a somewhat nearer future. But whatever the future sources of electrical energy production will be, two things are certain: the mix will be different from what it is now and there will have to be a much larger part of production with low carbon-dioxide emission.

This unpredictability holds at a global level, at a national level, and also at a local level. Production and consumption may change very quickly, faster than the network can be adapted. Building new lines can take several years; cables are in general faster, but not always. Demands on continuity of supply and voltage quality are increasing at the same time. All this calls for alternative ways of designing and operating the grid. These alternative ways are collectively known under the name "smart grid". We will briefly describe some of the possible solutions in Section 1.5 and in much more detail in Chapter 3 through Chapter 5.

1.5 THE SOLUTIONS

The classical way for the grid to address new challenges (like increase in consumption) has been to build new lines, transformers, etc. This has certainly been a very effective method of addressing the challenge, and in the past, it was also often the most cost-effective way. With the availability of new technology it is however certainly worth considering if this new technology can give the same result at a lower cost. Several of the solutions based on new technology will be explained in detail in the remainder of this book. Some of those solutions are natural continuations or modern versions of existing methods and technology whereas others offer completely new approaches. The terms "evolutionary" and "revolutionary" are sometimes used to distinguish between solutions that are a continuation of existing practices and those that are something completely new.

In this book, we will divide the different solutions into three groups:

• Solutions that only involve the electricity network

• Solutions that involved participation of network users, but where the network operator remains in control

• Solutions that are based on market principles: network users are given incentives to support the network.

Solutions that only involve the electricity network are discussed in Chapter 3. This includes the classical solution of building new lines, in

Section 3.1. Modern solutions aim at increasing the transport capacity at transmission level without having to build new lines, or with lesser lines than would have been needed otherwise. Several such solutions, HVDC, FACTS, dynamic line rating, and risk-based operational risk assessment, are the subject of Section 3.2. Whereas HVDC and FACTS are based on advanced power-electronic control techniques, dynamic line rating and risk-based operational risk assessment use a transport capacity and operating reserve that varies with time. The transport capacity and operating reserve are no longer based on the worst case during the year or during a season but on the actual weather circumstances of the hour. Using HVDC and FACTS, the worst-case transfer capacity is increased; this is not possible with the other methods. However, the other methods allow for the use of the actual available capacity at any moment in time.

A related solution that is moving to the forefront a lot recently is building a large transmission network allowing large scale transports over large distances, for example, between the east-coast and the west-coast of the United States or between Northern and Southern Europe. This could either be an HVDC network or an ac network at a voltage level of 750 to 1000kv. More about this in Section 3.3.

At distribution level, other solutions are the subject of a lot of research; we will discuss some of them in Section 3.5. Those solutions are largely based on using communication and automatic control much more than before. Storage, another subject that gets a lot of attention in the smart-grid literature, breaks with the old rule that electricity cannot be stored. A surplus of electricity can be stored until a shortage occurs. Storage as part of solutions in the grid is discussed in Section 3.4, but most of the applications of storage discussed involve in fact storage on customer side of the meter. This will allow the customer to participate more effectively in the various electricity markets.

In Chapter 4, a number of new solutions are described that involve the network users in the operation of the grid. With all these solutions, the network operator remains in control. The main trend to ensure a smooth introduction of new production in the grid is to set requirements on the production units (Section 4.1). Different requirements typically hold for distribution and transmission networks, but the general philosophy is that the introduction of new production should not require a change in design and operation of the grid. Alternative solutions based on intertrip (Section 4.3) and curtailment will allow more production to be connected without significant costs. Curtailment of consumption is currently only used in exceptional circumstances (Section 4.2), but in the future, curtailment may be more common (Section 4.4). Curtailment of production will allow for more new production to be connected and curtailment of consumption for more new consumption (Section 4.5).

Chapter 5 addresses solutions where control is shifted from the network operator to a market in which other network users are involved. This is a clear change compared to the existing method. With curtailment and intertrip, the network users get involved but the network operator remains in control and at any moment in time the network operator knows what the operational margins are. This is no longer the case with many of the market-based solutions that are being developed. Demand response (Section 5.3) is the market solution most discussed: by extending the hourly wholesale market to individual consumers they are giving incentive to reduce consumption whenever there is a shortage of production. Alternative methods are also discussed, where a higher price is only activated when there is a serious shortage. We will also discuss balancing markets (Section 5.4), network markets (Section 5.5) and ancillary-service markets (Section 5.6).

With the introduction of curtailment of consumption and participation of small customers in the various electricity markets, a lot of the development will have to take place on customer-side of the meter, i.e., beyond the control of the network operator or utility. We will address some of this in Section 5.7.

It is interesting to notice that many of the challenges posed to the electricity network are related to the opening of the market for electrical energy. The building and operation of the production units is no longer under the control of the network operator (utility) but is instead driven by market principles. An important discussion in the coming years is whether the solutions to these new challenges should involve additional markets or should be part of the natural monopoly owned by the network operator.

CHAPTER 2

The Challenges

This chapter will go into further details of the driving forces behind the shift towards smart grids: it will explain what are the actual challenges faced by the grid, the reasons for developing new technology. The first two challenges to be discussed have an immediate relation with the transition to a sustainable energy system: the need for integrating large amounts of renewable electricity production (to be discussed in Section 2.1) and the need for the integration of large amounts of energy-efficient equipment (to be discussed in Section 2.2). The opening of the electricity market has certain consequences for the grid that could form a challenge as well; this will be discussed in Section 2.3. Two of the original challenges faced by the grid for more than a hundred years now remain as actual as ever. Maintaining acceptable levels of voltage quality and continuity of supply is one of the primary aims of the grid. We will address this briefly in Section 2.4. Another primary aim, transporting electricity between production and consumption is the subject of Section 2.5, or more specifically the transport capacity needed to cope with the peaks in power transport. Finally, in Section 2.6, some of the indicators for quantifying the performance of the existing as well as the future grid will be presented.

2.1 RENEWABLE ELECTRICITY PRODUCTION

The integration of renewable electricity production in the grid is the main new challenge for many network operators, at distribution as well as at transmission level. A too large amount of renewable electricity production in the grid as a whole, or just locally, will endanger the reliability and quality of the supply to other network users. We say, in that case, that the "hosting capacity" of the grid is exceeded.

A lot has been said and written about the impact of renewable electricity production on the power system. Even for small amounts of new production, there is an impact. However, this impact is not always negative, and even when the impact is negative, it is not a concern as long as the performance of the electricity network remains within acceptable limits. The different impacts of renewable electricity production on the grid are described in detail in a number of textbooks [Ackermann, 2005, Bollen and Hassan, 2011, Dugan et al., 2003, Fox et al., 2007, Jenkins et al., 2000].

2.1.1 INTEGRATION IN THE DISTRIBUTION NETWORK

A large fraction of the new renewable electricity production is expected to be connected to the distribution network. In fact, the trend has started with wind power and solar power being connected exclusively to the distribution network. The terms "renewable electricity production" and "distributed generation" are often used as a synonym and as far as it concerns their impact on the distribution network, they are a synonym. In this section, we will use the term distributed generation so as to also cover the other emerging type of small-scale generation: combined-heat-and-power (see Section 2.2.4). The design of the distribution network does, in almost all cases, not consider the presence of production units. This does not immediately imply that it is not possible to connect production units to the distribution network. Connecting limited amounts will even result in an improvement of the grid performance as the production compensates for the consumption and in that way reduces the actual loading of the grid. Also, losses are in many cases lower with a limited amount of production. But above a certain amount of production, the performance of the distribution network deteriorates and would become unacceptable if no measures were to be taken.

The possible adverse impact of renewable electricity production on distribution networks can be summarized as follows.

- The injection of active power along a medium-voltage feeder or to a low-voltage network will reduce the voltage drop and may even result in an overvoltage. This is especially a concern for rural networks where the highest voltage levels are often already close to the maximum-permissible level even without distributed generation. Connecting a small amount of wind power or solar power can already result in unacceptable overvoltages. The main consequences for other network users that have been reported due to this are a significantly reduced lifelength of incandescent lamps, and tripping of solar panels by their overvoltage protection. See Section 3.5.2 for more details on voltage control in distribution networks.

- With larger amounts of distributed generation, uncontrolled island operation becomes a possibility. After the disconnection of a feeder (due to a fault or for maintenance purposes), the generators connected to the feeder may be able to supply the local consumption. This is an unwanted situation that may result in damage to equipment with other network users and dangerous situations for maintenance personnel. This should not be confused with "controlled islanding operation" as is used in industrial and commercial installations to improve reliability and that is discussed often as part of the developments of "microgrids" (see Section 3.5.3 and Section 5.7.7).

- The grid can become overloaded when the amount of local generation becomes bigger than the sum of maximum and minimum consumption. This situation is most likely to occur in rural areas with low consumption but large amounts of wind power and in suburban areas with large amounts of roof-top solar power.

- With larger amounts of certain types of production, the protection coordination can be endangered. This may either result in an incorrect trip or in a fail-to-trip situation. From the viewpoint of the other network users, both will result in an increased number of supply interruptions. A fail-to-trip event can also result in equipment damage, mainly in the network.

- Distributed generation could in some case result in an increase of the harmonic levels. More about that in Section 2.4.

2.1.2 INTEGRATION IN THE TRANSMISSION NETWORK

Compared to distribution networks, the integration of renewable electricity production in transmission networks is on one hand easier because the transmission network is built for having production connected, but on the other hand, the integration gets much more complicated because of the way in which the transmission system is operated. The latter is, in turn, strongly related to the high reliability requirements posed on the transmission system as well as a number of technical issues associated with the transport of large amounts of electrical energy over long distances.

Some of the impacts of renewable electricity production on the transmission system are summarized below. The term transmission system also includes in this case the subtransmission network (at voltage levels from about 50kv) and the interaction between the transmission network and the production units. A lot of the studies have been done for large wind-power installations connected to subtransmission networks, but several of the conclusions also hold for other types of renewable electricity production. Large amounts of distributed generation will have similar effects as large renewable electricity production is directly connected to the transmission system.

- The connection of new production units to the transmission system results in new power flows that can result in an overload or insufficient operating reserves (see Section 2.5 for further discussion on operating reserves). The need to keep operating reserves even for the worst-case situation further limits the hosting capacity. The worst case can be maximum production together with either maximum or minimum consumption. A serious challenge is the uncertainty in the type and location of future production together with the long time it takes to build new transmission lines.

- Strongly related with the first point is that large renewable electricity production units (large wind parks, new large hydropower installations) are often located in less-populated parts of the country where the transmission grid is weak.

- Adding new production units impacts the power-system stability, which is, in itself, nothing new. But a serious challenge is that it is not very well known how future production units will impact the stability. This in turn makes that large operational margins are needed to guarantee a secure operation which further limits the power flows and the amount of new production that can be connected to the transmission system. A much-discussed subject concerns the so-called "fault-ride-through": the ability of production units to remain connected (and support the grid) during large disturbances associated with serious drops in voltage or frequency. The challenge gets even bigger here because of requirements on distributed generation set by the distribution network operator, to disconnect when voltage and/or frequency deviate too much from their nominal values.

- The amount of electricity produced from renewable sources depends on the momentary availability of the resource, not on the actual need for electricity at that moment. Planning of the transmission system used to take place merely for two extreme cases: maximum consumption and minimum consumption, where the amount of production would change in tune with the amount of consumption. With large amounts of renewable electricity production, there appear four extreme cases: minimum production and minimum consumption, minimum production and maximum consumption, maximum production and minimum consumption and maximum production and maximum production. For transmission systems covering large geographical areas, differences in weather within the service area will further increase the number of extreme cases to be considered.

- Renewable electricity production is more variable and less predictable than conventional electricity production (fossil fuel, nuclear, and large hydro). This requires additional operational margins and reserves to maintain the same level of operational security. We will come back to operational security and the need to keep reserves in Section 2.5. Forecasting of production will be discussed further in Section 2.1.4.

- The shift from large convention production to renewable electricity production and distributed generation makes that during certain periods the amount of conventional units connected to the grid is insufficient. These units provide the ancillary services that maintain the transmission system secure and stable. We will discuss this further in Section 2.1.3.

- Large amounts of renewable electricity production could also require a rescheduling of the base-load production (see Section 2.2.1).

There is a rather important difference between the impact of renewable electricity production on distribution networks and on transmission networks. At distribution level, the impact on the network users becomes visible in the form of reduced reliability and voltage quality. This is however not the case at transmission level. The transmission-system operator has the duty to maintain sufficient operational reserve, and by doing that, a high level of reliability is maintained. To maintain sufficient operational reserve, the operator may however block certain transactions on the electricity market, for example, by rescheduling production or by curtailing the production from large wind-power installations. The consequences for the network users are a higher electricity price, possibly a higher network tariff, and most likely also more electricity production from polluting sources. In the worst case, rotating interruptions can be ordered. Even when we talk about reduced reliability in the long term due to a shortage of conventional generators (see Section 2.1.3), this means an increased use of rotating interruptions in order to maintain sufficient operating reserve.

2.1.3 REPLACING CONVENTIONAL GENERATORS

Large amounts of energy coming from renewable sources will reduce the need for conventional generators. The marginal costs of renewable electricity production is zero (sometimes even negative due to feed-in tariffs or other support schemes), whereas thermal power stations always have some marginal costs. On an open electricity market, renewable energy will therefore always be chosen first and energy from thermal units last. There are some complexities here, like the long start-up time of especially nuclear power units and the bidding behavior of hydro units on the electricity market, but that does not impact the overall trend.

The replacement of power from thermal units by renewable sources is obviously the aim of introducing renewables and a good thing. This replacement will however introduce a challenge for the operation of the power system. The reason is that large production units (thermal and hydro) not only produce electrical energy but also keep the power system stable and secure. Using existing technology and operational tools, stable operation of the power system will not be possible without at least some of those large production units. The large production units provide so-called "ancillary services" to the grid. Examples of ancillary services are the following (see also the discussion on ancillary-service markets in Section 5.6):

- **Operational reserve.** Part of the production capacity is kept in reserve so as to cope with unexpected shortages, e.g., due to the loss of a large amount of production, the loss of an important transmission line, or large prediction errors.

- **Frequency control.** Many of the large production units are equipped with power-frequency control. It is this control system that keeps the actual physical balance between production and consumption, in the complete interconnected system (the "primary control") and for each control area, typically the service area of a transmission-system operator (the "secondary control"). As a consequence, the control system also keeps the frequency close to its nominal value of 50 or 60 Hz. (Another way of looking at it is that the control system keeps the frequency constant and that the balance between production and consumption is a consequence of that.)

- **Voltage control.** Another control system maintains the voltage magnitude close to its required value throughout the transmission system. Voltage and reactive power are closely linked in the same way as (active) power and frequency are linked. There should thus be reserves in reactive power to allow voltage control even in case something unexpected happens.

- **Short-circuit capacity.** The short-circuit capacity is a measure for the amplitude of the current that will flow when a fault occurs in the transmission system. These currents can be very high and cause serious damage to equipment. The short-circuit capacity is however also a measure for the strength of the electricity network and for its ability to cope with changes in consumption. A too low value of the short-circuit capacity will have a number of adverse consequences, like risk of incorrect operation of the protection, risk of system instability after a fault or the loss of a component, insufficient voltage quality. The only source of short-circuit capacity at transmission level are the large conventional power stations. The less such units are in operation, the lower the short-circuit capacity.

- **Power-system stability.** The presence of more large production units in the transmission system in most cases makes the system more stable. Their contributions to voltage and frequency control and to short-circuit capacity were already mentioned. Also, their contribution to the total inertia of the system is important to prevent frequency and angular instability. Several large production units are also equipped with additional controllers to damp interarea oscillations.

The transmission system operator, being responsible for the secure operation of the system, is allowed to allocate "must-run production" that is essential for this secure operation. The transmission system operators also set limits to transport capacity between countries, between regions within a country, or between neighboring operators. The way in which this is implemented and how it impacts the market depends on the local market rules (see also Section 5.2). In general, the consequences are, however, that the market price will increase and that not all low-emission production can be scheduled. The reliability will not be impacted by all this at the short term, but in the long run, this could change.

An increased amount of electricity from renewable sources and the associated reduction of electricity from thermal units will reduce the income for the owners of the thermal units. At a certain moment, it is no longer attractive to keep those units and it becomes unattractive to build new units. This is in itself not a concern and the normal way in which market mechanisms make that new, more-efficient, technology replaces old technology. The production capacity from renewable production does however vary a lot with time and despite a high amount of installed capacity, there may be periods when the actual capacity is low. At that moment, the thermal units will be needed to supply the consumption. But those units may no longer be available at that moment. The result will be a shortage of production capacity with rotating interruptions as a consequence.

2.1.4 FORECASTING

The variability of production capacity from renewable sources is often mentioned as a serious disadvantage. This is only part of the picture. At distribution level, the variability is not much of a concern (with some minor exceptions), and even at transmission level, the concern is mainly limited to the long-term planning of production units. What matters more is the difficulty in predicting the variations in production on a time scale from a few minutes up to a few days. For small amounts of renewable electricity production, the prediction errors in consumption are dominating. These are taken care of by the "load-following reserves" that are part of the power-frequency control, which can also take care of the prediction errors in production.

For large amounts of renewable electricity production, additional reserves are needed because of the large prediction errors that can occur. The production errors for wind-power production in North-Eastern Germany (a geographical area of about 300 by 450 km) during 2009 are presented here as an example. The distribution of the production error in shown in Figure 2.1. The maximum production from wind power during 2009 was somewhat over 8000MW.

The probability of more than 1000MW error is about 3%; the probability of more than 2000MW error (a quarter of maximum production) is still 0.3%, corresponding to 26 hours of the year. What matters for the planning of the operational reserve is not only the probability distribution but also the worst-case situation. The largest production error during 2009 is shown in Figure 2.2. The plot on the left shows the predicted and the actual production. The prediction data are the ones used for the day-ahead market based on weather prediction obtained from the weather office. The actual production is estimated from measurements at a number of selected locations.

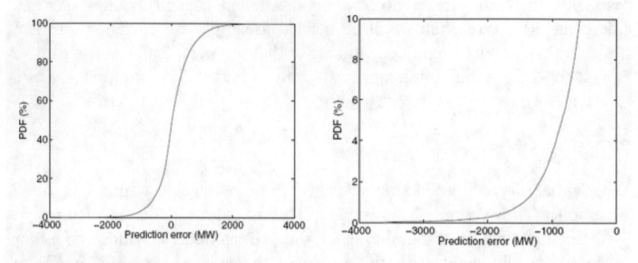

Figure 2.1: Probability distribution function of the prediction error for wind power: full scale (left) and details for production shortages (right).

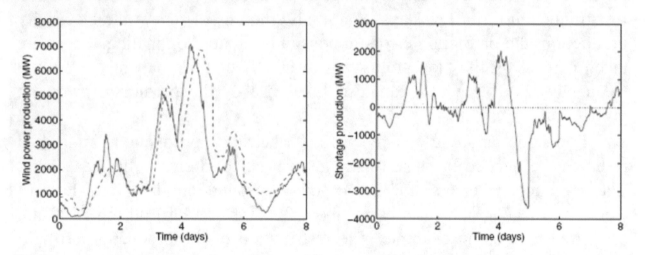

Figure 2.2: Actual (left; blue solid line) and predicted (left; red dashed line) wind-power production and prediction error (right).

Because of the arrival of a weather system a few hours earlier than predicted, there are huge deviations between the predicted and actual amount of production from wind power. The result is the large production error of 3600MW shown in the right-hand side of the figure. The drop in wind-power production will have to be compensated by the increase in production from other sources or by import from other areas. But limitations in transport capacity into the area and long ramping times of thermal power stations can make that a local shortage of production occurs. The transmission-system operator has no other choice in that case than to curtail local consumption by means of rotating interruptions.

2.2 ENERGY EFFICIENCY

Reducing energy consumption is the most effective way of reducing greenhouse gas emission, and all the other adverse impacts of energy consumption. This can be achieved in a lot of different ways, some of which are related to the electric power system some of which are not. Here we will only discuss those that have an impact on the electric power system: reduced electricity consumption, reduction in network losses, joined production of electricity and heat in the same installation and a shift from other sources of energy to electricity as an energy carrier. It is the latter that is expected to have the biggest impact, but even the others require at least a brief discussion here. In this section, it is important to realize that a reduction in total energy consumption by society is generally expected to result in an increase in electricity consumption.

2.2.1 REDUCED ELECTRICITY CONSUMPTION

The electricity consumption can be reduced in two ways: by using less appliances and using appliances less often, or by using appliances that consume less electricity. The overall reduction in the consumption of electrical energy will be advantageous to the electric power system; it will likely reduce peak load (more about this in Section 2.5), reduce losses and improve the stability. But even something as simple as reducing electricity consumption could pose a challenge to the grid. When the overall reduction in consumption results in a large reduction of the minimum consumption, the scheduling of production units will need to be adapted. In most countries, so-called base-load production is running continuously for weeks or even months. This is relatively cheap production (nuclear and coal are often used for this) that takes a long time to start. The amount of base-load production can obviously not be more than the minimum consumption. A large reduction in the minimum consumption thus means that less base-load production can be used and that more expensive peak-load production has to be operated. When minimum consumption drops more than maximum consumption, it could even result in a shortage of peak-load production with high electricity prices or rotating interruptions as a result.

The general trend is however not towards a reduction in electricity consumption. Instead, more and more appliances are being used, but they can do the same job while using less electricity each, but the total consumption still shows an increasing trend. The efficiency of household appliances has improved a lot and so has the efficiency of electrical-motor drives. The most recent trend is the phasing out of incandescent lamp (in many countries pushed by legislation) and their replacement by more efficient lamp types. The increase in efficiency (sometimes up to 90%) is made possible by the availability of cheap and effective power-electronic converters. These power-electronic converters do however again pose a potential challenge for the grid. The current waveform taken by a converter is much more non-sinusoidal as the current taken by an electrical motor without power electronics or by an incandescent lamp. The non-sinusoidal currents (called "harmonic currents") could endanger the voltage quality experienced by other network users; they could also result in reduction of life for components in the grid. It is possible to produce devices that inject reduced amounts of harmonic currents, but this would make the devices more expensive and bigger, reduce the efficiency of the device, increase the amount of electronic waste and probably also reduce the life of the device. All this is still being studied and discussed in standard-setting organizations and no agreement has been reached yet. This increase in harmonic current together with a reduction in active power is a new situation for the grid (see Section 2.4).

2.2.2 LOSSES IN THE GRID

There are losses associated with any transport of electricity through the grid. When sending a current I through a component with resistance R, the losses are equal to $I^2 \times R$. This relation is one of the reasons for using high voltage to transport electric power over long distances. The higher the voltage, the lower the current and thus the lower the losses for the same amount of power.

The losses in the grid are between 5 and 10% of the production for most countries, depending on the consumer density and the distance between production and consumption. Losses in the grid can thus not be neglected. However, the losses in the transport of electricity are still very small compared to the losses in production and in certain types of consumption. The efficiency of thermal production units is between 40 and 60%. That means that the amount of primary energy is 1.7 to 2.5 times the amount of electricity produced. Also on the consumption side, much can still be gained. About 15% of electricity is used for lighting purposes. Using low-energy lamps, this can easily be reduced by a factor of four, resulting in a reduction in total consumption by 11%, more than the total losses in the grid. The total potential for savings is thus much higher on production and consumption side than in the grid, but it is still worth to look at ways to reduce the losses in the grid.

As mentioned before, the losses are proportional to the square of the current, the resistance and the distance. Using higher operational voltage will reduce the losses. An increase in operational voltage by 5% will reduce the losses by 9.3% because of the square relation.

The losses are proportional to the distance over which the electricity is consumed. Distributed generation will therefore generally result in a reduction in losses (as long as production is less than twice the consumption). Also, building large production units generally closer to consumption centers will reduce the losses. If the electricity is produced close to the consumption, the losses are much less than when the electricity has to be transported over a long distance. According to National Grid (the British TSO), the effectiveness of producing electricity in the north of Scotland is about 20% less than producing it in the south of England. (Note that this does not mean that the losses are at 20%; adding production in the north will increase the losses; adding production in the south will reduce the losses. In total, the difference is 20% of the produced power.)

The fact that the losses are the square of the current makes that a reduction in current is the most effective way of reducing the losses. A general reduction in consumption will also reduce the losses. It further pays to keep the consumption constant. The less the variation in current, the less the losses as percentage of the total transported energy. (In mathematical terms, the losses are proportional to the square of the average current plus the square of the standard deviation.)

Reducing the resistance also reduces the losses. The most effective method is to use superconducting material, for which the resistance is zero. Using superconductors for long distance transmission would be extremely expensive and the cooling required could well consume more electricity than is saved from the losses. A moderate reduction in resistance seems more reasonable, this can be done by increasing the cross-section of conductors and by using low-loss material in transformer cores.

2.2.3 INCREASED ELECTRICITY APPLICATIONS

As mentioned before, a reduction in energy consumption can be achieved by increasing the electricity consumption. This is partly because electricity is such an efficient energy carrier, but also because the production mix is shifting more towards low-emission sources like renewable electricity production. Space heating using direct electric heating with electricity produced by low-efficiency coal-fired power plants will give more carbon-dioxide emission than gas heating, but using heat pumps and a high amount of hydro and renewables in the electricity production mix will certainly give less emission than gas heating.

An efficient gasoline car will result in less emission than an electric car powered from a low-efficiency coal-fired plant, but using large amounts of solar and hydro will reduce overall emission when shifting to electric cars.

These two examples show that the shift from direct use of fossil fuel to electricity as an energy carrier will only be effective if it goes hand in hand with an increased amount of electricity coming from renewable sources. Building new coal-fired plants to power electric cars makes no sense.

The impact on the grid of the shift to electric heating and electric vehicles will be an increase in peak consumption and increased losses in the grid. The shift to electric vehicles is expected to go relatively slow (the one million electric vehicles envisaged for the United States in 2015 would only increase peak consumption with 1% [Ungar and Fell, 2010]), but a shift to used in some countries) and could increase the peak load by a factor of two or more.

The consequences of any increased consumption are not unlike the consequences of additional production. The difference is that with additional consumption there is in most cases no production to be compensated so that some of the adverse effects become noticeable already for small amounts of increased consumption. In other words: the hosting capacity for new consumption will often be smaller than the hosting capacity for new production. A brief summary of the consequences of increased production is given in the bullet points below.

- The voltage drop will increase resulting in undervoltages. This will happen first in rural networks where the voltage drop is the limiting factor. Domestic customers in the more remote areas typically have no access to district heating or gas for heating so that they are likely to be the first to switch to electric heating.

- Increased consumption will obviously increase the risk of overloading, in distribution as well as in transmission networks. A large increase in consumption over a large geographical area could even result in a shortage of production capacity with the before-mentioned rotating interruptions as a consequence.

- It will initially be difficult to accurately predict the consumption as a function of time for electric vehicles and electric heating, so that large prediction errors will be likely. To compensate for the prediction errors, more operating reserve is needed. This will reduce the secure transport capacity and the amount of generation available to cover peak consumption.

2.2.4 COMBINED-HEAT-AND-POWER

A third trend related to increased energy efficiency could also pose a challenge to the grid. That is the use of combined-heat-and-power (CHP, also known as "co-generation") to produce heat and electricity at the same time. The impact of CHP on the grid is very similar to the impact of renewable electricity production (Section 2.1) with as a main exception that the production by CHP is much stronger correlated with the consumption that the production by renewable energy. This is illustrated in Figure 2.3 for Western Denmark using hourly production and consumption data for 2008.

During periods of high electricity consumption, CHP can be relied on; when the consumption increases above 3000MW, the production from combined-heat-and-power increases quickly. For wind power, there is no such relation. Although there is a small positive correlation (the correlation coefficient is 0.2), wind power cannot be relied on to provide power during peak consumption. The presence of CHP reduces the need for production capacity from other sources by about 1000MW whereas this reduction due to the presence of wind power is small. For a quantification of the usefulness of different types of production, the so-called "capacity credit" needs to be calculated for wind power as well as for CHP. The capacity credit is the amount of conventional production that can be removed without reducing reliability when adding a certain amount of new production.

2.3 THE ELECTRICITY MARKET

The restructuring of the electricity market, which started around 1990 in a small number of countries at about the same time, has impacted the grid in a number of ways. Further development of the electricity market is expected to have further impact.

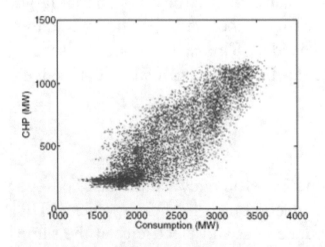
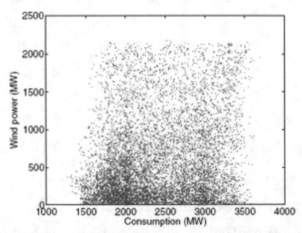

Figure 2.3: Correlation between consumption and production from combined-heat-and-power (left) and production from wind power (right).

As we will see in more detail in Section 5.1, the electricity market consists of a number of more or less separated markets, the wholesale market, the retail market and the balancing or real-time market are the ones that have the main impact on the grid. There are also a number of commodity markets where products like futures, forwards and options are traded. These do not have any impact on the grid and we will not discuss them further here.

The main impact on the transmission grid is due to the day-ahead wholesale market. This is where it is decided which production facilities will cover the expected consumption. The rules of the wholesale market determine which production is scheduled, no longer the interests of the transmission-system operator. The transmission-system operator however still has the responsibility to maintain a high operational security. To prevent the market from endangering this high operational security the transmission-system operator defines secure transport capacity and essential production. The market will next be adjusted such that these constraints are fulfilled. The market cannot result in an insecure operational state; it is the transmission network that sets barriers to the free operation of the market.

In the future, it is expected that the geographical extent of wholesale markets will grow further. The Nordic wholesale market operated by NordPool already covers five countries in Northern Europe with an area of 2000 by 1500 km. Market coupling is in place with the markets in Poland, Germany and The Netherlands, and by 2015, the day-ahead wholesale markets should be coupled throughout most of the European continent. This will result in a single electricity market covering more than 25 countries and 500 million people. The planned merger of PJM and the Midwest independent transmission system operator (MISO) will create a single wholesale market covering 23 U.S. states, Washington D.C., and one Canadian province and form an even bigger geographical area with a much higher consumption. The bigger the geographical area covered by the market, the bigger the need for transport of large amounts of power over long distances.

The retail market has much less impact on the operation of the grid. A retailer purchases electricity on the wholesale market and sells this to its customers. But this is in fact only a financial transaction; there is no physical transaction associated with this. If a consumer switches to another retailer, this does not impact any of the power flows in the grid. It is, however, the retail market where the price elasticity of the consumption comes in. The retailer estimates how much its customers will consume each hour; these values are the basis for reaching a market settlement. The system operator will have to balance any difference between the estimated and the actual production. This is taken care of by the balancing or the real-time market. The deviations and the balancing of them will also result in further power flows with the associated risk of insecure operation.

Currently, not many consumers are exposed to the price variations on the wholesale market. Therefore, the consumption is rarely adjusted to the electricity price; in economic terms, the price-elasticity is low. In the future it is expected that retailers will offer hourly prices to their customers. The price elasticity of the customers is very much unknown at the moment. During the introduction stage of hourly prices to many consumers, the differences between predicted and actual consumption are expected to become larger. The economic consequence will be large price variations on the balancing and real-time markets. For the grid, it means more unpredictable power flows and the need for keeping larger reserve margins to be able to cope with larger prediction errors.

The wholesale market does not yet have any impact on the distribution network. But when large amounts of distributed generation become part of the wholesale market, it could impact the distribution network. Demand response could also result in unexpected power flows through the distribution network. At distribution level, the maximum power flows are mainly what matters, so the additional uncertainty introduced by the markets is less of a concern.

Here it should be noted that similar market principles that result in congestion in the grid can also be used to support the grid. We will come back to this in further detail in Chapter 5.

2.4 CONTINUITY OF SUPPLY AND VOLTAGE QUALITY

2.4.1 NEED FOR IMPROVED PERFORMANCE

Continuity of supply and voltage quality belong to the primary aims of the power system. The design of the network is such that the continuity of supply and the voltage quality are sufficient for the majority of network users. In the literature, a distinction is often made between "continuity of supply" (or, "reliability") and "voltage quality" (for which also the term "power quality" is used). There is no natural distinction point between them, but roughly speaking, continuity of supply considers the absence of supply interruptions whereas voltage quality considers all other deviations from the ideal voltage. Phenomena like voltage dips, harmonics and flicker are counted as part of voltage quality; long and short interruptions are considered part of continuity of supply. More on this can be found in the various books, reports and papers on this subject [Baggini, 2008, Bollen, 2000, Bollen and Gu, 2006, Caramia et al., 2009, CEER, 2008, Dugan et al., 2003, Schlabbach et al., 2001].

Maintaining sufficient levels of continuity of supply and voltage quality has been part of the design of electricity networks since its very beginning. The result of the over 100 years of development in this has resulted in a continuity and quality that are sufficient for most network users most of the time. But there remain occasions when the continuity or quality is insufficient. The actual challenges posed to the grid at the moment can be summarized as follows:

- Situations where many network users are without electricity for a long time should be avoided. This can be due to an incident in the transmission system resulting in a blackout or due to many failures in the distribution system during severe weather or natural disasters. Blackouts are generally considered as unacceptable; loss of electricity during natural disasters (hurricanes, severe earthquakes) on the other hand is seen as unavoidable. The border between a natural disaster (when a black-out is acceptable) and extreme weather (when a black-out is not acceptable) is shifting upwards. This puts higher and higher demands on the grid to tolerate more extreme external circumstances.

- A gradual increase of the average continuity of supply and quality is demanded from the network operator; for example, the average number of interruptions per customer per year (SAIFI) and related reliability indices should show a gradual reduction.

- For certain network users, the existing continuity of supply and voltage quality is insufficient. This is summarized in Figure 2.4. In some cases, investments need to be made by the network operator (when the existing continuity and quality is too low); in other cases, the network user has to stand for the investments (when its requirements are higher than for normal customers). The main issue here is what should be seen as normal continuity and quality. Defining this is a challenge by itself.

		Continuity of supply and voltage quality		
		Low	Normal	High
Requirements of network user	Low			
	Normal	Investments by network operator		
	High	Investments by both	Investments by network user	

Figure 2.4: Need for investments when the existing continuity and quality do not fulfill the requirements of the network user.

An important role of the regulator in continuity of supply and voltage quality is to define the border between "normal" and "low" in Figure 2.4. When the performance of the network is below this limit, it is the network operator that has to make the investments. When the performance is above this limit, but still insufficient for the network user, it is the network user that makes the investments.

2.4.2 NEW PRODUCTION AND CONSUMPTION

The voltage quality is impacted by new production and consumption in a number of ways. Both will cause additional emission which could deteriorate the voltage quality. The emission from distributed generation is discussed frequently (see Bollen and Hassan [2011, Chapter 6] and Dugan et al. [2003] for overviews), but serious problems due to this emission are only expected at a small number of locations. Also, energy-efficient equipment (adjustable-speed drives, compact fluorescent lamps) has a higher emission level than its traditional (less energy efficient) counterpart.

The impact of distributed generation on harmonics requires some clarification. The actual harmonic levels from distributed generation are small and much smaller than the harmonic levels coming from typical low-voltage equipment like televisions or computers. There are however two aspects of this that have to be explained. The first is that the power-electronic converters used for the connection of distributed generation do not inject the same kind of harmonic distortion to the grid as other equipment. Next to the "normal harmonics", these converters inject even harmonics, interharmonics, and high-frequency harmonics. Even harmonics are frequency components at even multiples of the power-system frequency; interharmonics are frequency components that are not an integer multiple of the power-system frequency; high-frequency harmonics are frequency components in the range from 2 to 150 kHz. Because these "new harmonics" were absent from the grid before, acceptable limits ("planning levels", "compatibility levels" and "voltage characteristics") have been chosen very low. With the introduction of new types of converters, these levels can be exceeded very quickly. That does not however imply that there are any problems associated with this.

The second aspect has to do with the available disturbance allocation. Normally, when adding equipment to the grid, an increase in harmonic level is associated with an increase in active-power consumption. The latter requires strengthening of the grid. This strengthening also takes care of the increase in harmonic distortion as long as the harmonic currents are not excessive. This, in turn, is taken care of in harmonic standards. All this no longer holds with the introduction of production units that inject harmonics. Suddenly, an increase in harmonics comes with a reduction in active power. There is no need to strengthen the grid, so the increased level of harmonics is not taken care of either. The same is the case when equipment is replaced by more efficient equipment taking less active power but injecting more harmonic current.

New types of production will also impact the voltage quality in two other ways. Capacitor banks and long underground cables are a common characteristic of new production installations. These result in harmonic resonances that can give high levels of harmonic voltage and currents. The second impact is related to the reduction in the number of conventional production units connected to the transmission system. This makes the transmission system weaker and gives a higher level of harmonic voltage distortion. Also the number of voltage dips will somewhat increase.

2.4.3 RELIABILITY INDICATORS

Reliability indicators have been used in many countries for many years now. The most well-known ones are SAIFI (System Average Interruption Frequency Index) and SAIDI (System Average Interruption Duration Index). Both are defined in a consistent way by IEEE Std.1366 [CEER, 2008, IEEE, 2003a], together with many other reliability indices.

An important trend in some countries is to not only consider average interruption frequency (SAIFI) and average unavailability (SAIDI) but to consider also the values for the worst-served customers. This can be done in two complimentary ways. A threshold can be set on what are the highest values of interruption frequency and unavailability that would be acceptable. (This corresponds to the border between "normal" and "low" quality and continuity in Figure 2.4.) The number or percentage of customers for which this level is exceeded is next used as an indicator.

An alternative approach is to consider a small percentage of "worst-served customers", for example, 5%. The lowest interruption frequency for these 5% worst-served customers could be used as an indicator. Thus, 95% of customers have an interruption frequency that is better than this. The same can be done for the unavailability (number of minutes per year without electricity).

Two types of interruptions, related with smart grids, that need to be considered in performance indicators are "curtailment" and "demand care of this because there is a wide range of curtailment and demand response and because it remains unclear how this will develop in the future. When including curtailment in reliability indicators one has to, one way or the other, include the inconvenience that the curtailment has for the customer. Turning off a refrigerator for 1 hour will have almost no impact, but turning off the light for 10 minutes could be a significant inconvenience. When curtailing heating or cooling, the outside temperature might also be considered as an indicator. With demand response, the situation gets even more complicated because it is so much harder to measure.

2.4.4 VOLTAGE QUALITY INDICATORS

Voltage quality indicators are less commonly used than reliability indicators, but some guidance exists, for example, in the form of the European voltage-characteristics standard EN 50160 [CENELEC, 2010]. Voltage quality indicators have to be divided into two groups: those related to voltage-quality events (at the moment: voltage dips and rapid voltage changes; in the future: possibly also transients) and those related to voltage-quality variations (at the moment: voltage-magnitude variations, harmonics, flicker and unbalance).

Voltage-quality events can be counted: indicators for those phenomena give basically a number of events or an average number of events per customer. They are very similar to indicators for interruption frequency (SAIFI). Where it concerns voltage dips typically an average number of events per measurement location or per customer is given for dips of different duration and residual voltage. According to the recent proposals by an international working group, it is also relevant to give the number of dips for the worst-served customers [CIGRE, 2010].

Where it concerns indicators for harmonics and other voltage-quality variations, another approach is needed. The average value during, for example, a year is not an appropriate indicator. Instead, a maximum-acceptable level should be defined. Suitable indicators could be the number of times per year or the number of hours per year that these limits are exceeded. For most variations, it is reasonably well understood which levels are acceptable, although discussion remains ongoing on some details. For other variations, there is no good indication on what is acceptable, among others due to some recent developments.

The limits for fast fluctuations in voltage magnitude are based on the visibility of flicker and other light-intensity variations with a standard incandescent lamp. Due to the phasing out of these lamps the appropriateness of these limits is currently under discussion. As different types of non-incandescent lamps all react differently to voltage fluctuations, it will no longer be possible to define a "standard lamp".

With voltage harmonics, some of the limits were based on the existing levels in the network about 30 years ago. The levels of even harmonics (even multiples of the power-system frequency) and interharmonics (non-integer multiples of the power-system frequency) were very low, with very low limits as a result. Several types of modern equipment (including wind turbines) causes distortion at these frequencies and it might be very difficult to comply with these low limits in the future [Yang et al., 2011].

For frequencies above 2 kHz, there are no limits at the moment, for the simple reason that there was no distortion above this frequency in the past. It was also rather difficult to measure these frequencies, and no problems were reported below 150 kHz anyway. Many modern appliances (including solar panels and some types of energy-efficient lighting) do cause distortion at these frequencies and work has recently started on defining suitable limits [Larsson et al., 2010].

2.5 TRANSPORT CAPACITY

2.5.1 DISTRIBUTION

A basic design criterion for the power system is that the demands of all network users are fulfilled at all time. This includes the peak demand. Let us first have a closer look at this where it concerns a local area, for example, a medium-voltage distribution network connected to the transmission grid. The classical situation (still the case almost everywhere) is where there is only consumption in the distribution network. The connection with the transmission grid should be able to transport the highest amount of consumption. This might be for the coldest winter day, for the hottest summer day, or for any other situation that would cause the highest amount of consumption.

But that is not enough to guarantee a reliable supply. To build new lines or cables requires a few years, so the growth in consumption also has to be considered in deciding about the need for transport capacity. A grid component can also fail, and, even in that case, it should be possible to supply the consumers. An interruption of a few hours is often acceptable, but a longer one is not. As it might be difficult to repair the faulted component quick, most network operators maintain a certain reserve so that they can restore the supply within a few hours even if the repair of the faulted component would take much longer. The required transport capacity from the transmission grid into the medium-voltage grid should thus be at least equal to the sum of the following three terms:

• The highest consumption,

• The expected growth in consumption during the coming years,

• An amount of reserve to cover the loss of an important component.

This could be referred to as a "worst-case design": the grid has to be able to cope with the loss of an important component at the same time as the maximum consumption a few years into the future. The probability of this happening may not seem very large, but this is the way in which a high continuity of supply is ensured. With the increased use of distributed generation (see Section 2.1.1 and Section 2.2.4), the demands on the distribution network change as well. Instead of only having to consider "maximum consumption" (highest consumption with minimum production), the design also has to consider "maximum production" (highest production with minimum consumption). The introduction of new production, together with new consumption, could make it especially difficult to predict the growth in production and consumption.

2.5.2 SUBTRANSMISSION AND TRANSMISSION

At subtransmission and transmission level, the situation is very similar. The presence of production is normal here and considered in the design. Next to future growth in consumption and production, the network also has to be able to cope with large transports of power due to the electricity market. There is again the uncertainty here about what the future levels of production, consumption and long-distance power transport will be. The need to cover all possible situations is somewhat less at transmission and subtransmission level than at distribution level. The transmission-system operator can block transactions on the electricity market when that would result in insecure operation of the grid. (See Section 2.3.) The transmission grid should, however, always be able to supply the local consumption. This means that the local production plus the import capacity from elsewhere should be enough to cover the local consumption. Using the same terminology as before, the import capacity from elsewhere (which is what the network operator can influence) should be at least equal to the sum of the following terms:

- The maximum consumption minus the minimum production,

- The expected growth in maximum consumption minus minimum production,

- A reserve margin.

The reserve margin is needed to guarantee sufficient continuity of supply even in the future. The amount of reserve required for this varies somewhat.

At the highest voltage levels, the operation of the system is always such that the loss of any component (line, transformer, large production unit, etc.) will not result in an interruption for any of the network users. Once this would no longer hold, typically after the loss of a major component, the network operator takes actions so that the requirement holds again within typically 15 to 60 minutes. This is called the (N-1) criterion, and it has been the basis for guaranteeing the extremely high reliability we have come to expect from the transmission system. When the (N-1) criterion holds, we say that the system operation is secure. The maximum amount of power to be transferred over a transmission line or, for example, between two parts of a country, for which the (N-1) criterion holds, is called the "secure transport capacity". It is this secure transport capacity that should not be exceeded and that is used when deciding if intervention in the electricity market is required.

When deciding about the need for building more transmission lines, the situation gets a bit more complicated. If one major component is out of operation for repair or maintenance, the operation should still be secure. There should thus be sufficient reserve even in that case. One might say that the design should be according to an (N-2) criterion. Such an (N-2) criterion is, in most cases, used at the highest voltage levels, but not at somewhat lower (subtransmission) levels.

2.5.3 PRODUCTION

At system level, where the production capacity for a whole country is considered, the need for reserves is even bigger. In the deregulated market, i.e., the open electricity market, there is no longer anybody responsible for planning production capacity. Still, many system operators do an assessment of the risk that there will be insufficient production capacity.

For example, the Swedish transmission-system operator publishes a report every autumn in which the available production capacity is compared with the expected maximum consumption during winter. It is assumed that only part of the impact capacity is available. For different types of production, different availability is considered: thermal and hydropower is for 90% available; wind power is for 6% available. This production and import should cover the "maximum consumption", which is in this case the estimated maximum consumption during a "ten-year winter" (28 200MW), plus 325MW primary reserve, plus 1200MW secondary reserve.

As a result of the different reserve margins, the amount of production needed is at least 17% more than the expected maximum consumption. The actual production capacity is exceeding this. The situation is very similar in other interconnected systems: the installed production capacity far exceeds the expected maximum consumption. The reason for having significantly more production capacity than the maximum consumption is not just the need for keeping reserves at production level. It is also because of limited transmission capacity between parts of an interconnected system. One of the reasons for enlarging electricity markets is to share reserves over a larger geographical area. This is of course only possible if there is enough transport capacity between different parts of the system.

2.5.4 INSUFFICIENT TRANSPORT CAPACITY

The consequences of insufficient transport capacity are different at different voltage levels. They can be summarized as follows:

- At distribution level, insufficient transport capacity will result in supply interruptions for the network users.

- At subtransmission level, it will result in temporary shutting down of production units, an increased risk of interruptions, and possibly a minor increase in electricity price.

- At transmission level, the consequences are large differences in electricity price between neighboring areas, higher emission of carbon dioxide and, in the worst case, rotating interruptions.

Several of the before-mentioned challenges result in an increased risk for insufficient transport capacity: new production from renewable sources or from combined-heat-and-power, new consumption from electric vehicles and electric heating and the further opening of the wholesale electricity market. Increasing the transport capacity is thus an important challenge for the grid at all voltage levels.

2.6 PERFORMANCE INDICATORS

Among others, for regulatory purposes, it is important to be able to quantify how well a power system (or, a network operator) performs its tasks. So-called performance indicators have been introduced already several years ago, especially for continuity of supply and voltage quality as a way to determine how well the network performs its aim, and to give incentives to network operators to make the right investment decisions. This so-called "incentive-based regulation" is likely to be extended to other indicators so as to give incentives to invest in cost-effective technology.

A list of 34 performance indicators has been proposed by the European Energy Regulators in their position paper on smart grids [ERGEG, 2010a] and has been taken over by several other European organizations since then. The proposed indicators, grouped by "benefit", are as follows:

1. Increased sustainability

    (a) Quantified reduction of carbon emissions
    (b) Environmental impact of electricity grid infrastructure

2. Adequate capacity of transmission and distribution grids for "collecting" and bringing electricity to consumers

    (a) Hosting capacity for distributed energy resources
    (b) Allowable maximum injection of power without congestion risks
    (c) Energy not withdrawn from renewable sources due to congestion and/or security risks

3. Adequate grid connection and access for all kind of grid users

    (a) first connection charges for generators, consumers and those that do both
    (b) grid tariffs for generators, consumers and those that do both
    (c) methods adopted to calculate charges and tariffs
    (d) time to connect a new user

4. Satisfactory levels of security and quality of supply

    (a) Ratio of reliably available generation capacity and peak demand
    (b) Share of electrical energy produced by renewable sources
    (c) Measured satisfaction of grid users for the "grid services" they receive
    (d) Power system stability performance
    (e) Duration and frequency of interruptions per customer

(f) Voltage quality performance

5. Enhanced efficiency and better service in electricity supply and grid operation

(a) Level of losses in transmission and in distribution networks (absolute or percentage)

(b) Ratio between minimum and maximum electricity demand within a defined time period (e.g., one day, one week)

(c) Percentage utilization (i.e., average loading) of electricity grid elements

(d) Availability of network components (related to planned and unplanned maintenance) and its impact on network performances

(e) Actual availability of network capacity with respect to its standard value (e.g., net transport capacity in transmission grids, DER hosting capacity in distribution grids)

6. Effective support of transnational electricity markets

(a) Ratio between interconnection capacity of one country/region and its electricity demand

(b) Exploitation of interconnection capacity (ratio between mono-directional energy transfers and net transport capacity), particularly related to maximization of capacity according to the regulation on electricity cross-border exchanges and the congestion management guidelines

(c) Congestion rents across interconnections

7. Coordinated grid development through common European, regional and local grid planning to optimize transmission grid infrastructure

(a) impact of congestion on outcomes and prices of national/regional markets

(b) societal benefit/cost ratio of a proposed infrastructure investment

(c) overall welfare increase, i.e., always running the cheapest generators to supply the actual demand

(d) Time for licensing/authorisation of a new electricity transmission infrastructure

(e) Time for construction (i.e., after authorization) of a new electricity transmission infrastructure

8. Enhanced customer awareness and participation in the market by new players

(a) Demand side participation in electricity markets and in energy efficiency measures

(b) Percentage of consumers on time-of-use / critical peak / hourly pricing

(c) Measured modifications of electricity consumption patterns after new pricing schemes

(d) Percentage of users available to behave as interruptible load

(e) Percentage of load demand participating in market-like schemes for demand flexibility

(f) Percentage participation of users connected to lower voltage levels to ancillary services

These indicators are supposed to be used by national regulatory authorities as a way of assessing the performance of network operators. The same indicators could however also be used to assess research and development and demonstration projects on smart grids. Network operators could use this list to assist in decision making on investments and in future development. More recently, a status report has been issued by the European Energy Regulators [CEER, 2011] in which, among others, the use of these indicators in various European countries is mapped.

Some of these indicators are already in use in many European countries. Most progress has been made on reliability indices ("Duration and frequency of interruptions per customer") but also other indicators are in use in several countries.

CHAPTER 3

Solutions in the Grid

The different challenges presented in the previous chapter can be addressed in different ways. In this chapter, we will present some of the solutions that only involve the electricity network. Classical solutions will be addressed in Section 3.1; methods for increasing the transport capacity, especially in the subtransmission and transmission networks, are addressed in Section 3.2. Two specific solutions, large transmission networks and storage, are the subject of Section 3.3 and Section 3.4, respectively. Section 3.5 will discuss different solutions for modernizing the distribution network, including protection, voltage control, microgrids and automatic supply restoration. Finally, Section 3.6 presents some applications that make use of the large amount of measurement data that comes available from other smart-grid solutions.

3.1 CLASSICAL SOLUTIONS

The default classical solution to most of the challenges mentioned in Chapter 2 is to add more primary components to the network: lines, cables, transformers and substations. This is often the most effective method, and for many years, this has also been the most cost-effective solution. We will only give a brief overview here of the classical solutions in relation to the various challenges. For more details, the reader is referred to the many books on power-system design, including the classical texts by Gönen [1986, 1988].

At distribution level, building new feeders and strengthening existing feeders can solve a lot of problems: the voltage drops and rises will be less, more power can be transferred before the feeder gets overloaded, and even the level of some voltage-quality disturbances becomes less. Strengthening existing feeders means, in this case, increasing the cross-section of the wires. Wires with a bigger cross-section can accept more current before they get overloaded, and for the same current, the voltage drop along the feeder is less. In terms of the challenges discussed before, building new feeders and strengthening existing feeders will allow more distributed generation to be connected, will allow more energy-efficient equipment to be connected, and will improve the voltage quality. A serious concern with distributed generation like small-scale wind and solar power is the voltage rise due to the injection of active power on a distribution feeder. Strengthening the feeder will reduce the voltage rise and thus the risk of overvoltages. Building dedicated feeders for wind-power installations is a common practice with some network operators. This also solves some of the protection-coordination problems that occur due to distributed generation.

With energy-efficient consumption like charging of electric cars and electric heating, the risk of undervoltages and overloading is the main concern. The risk can be reduced by strengthening the feeder. Also, the harmonic voltage distortion due to large numbers of energy-efficient lamps and adjustable-speed motor drives can be reduced in this way.

Replacing overhead lines by underground cables has a positive effect on reliability. Together with the replacement of the overhead lines, this also requires replacement of some of the switchgear. The reason for this is that the repair time of underground cables is much longer than of overhead lines. Failure of underground cables is much rarer than of overhead lines, but once they fail, it can take much longer to repair them. Therefore, a backup supply is needed as to prevent very-long interruptions for the customers.

The higher reliability of underground networks is not only due to the lower failure rate of cables. The increased switching possibilities were already mentioned in the previous paragraph. Next to that, failures in underground cables occur much more random than on overhead lines where the weather is the main factor. The failure of multiple underground cables close together in time is rare, whereas one storm can damage lots of overhead lines within a few hours. The restoration of the supply after a major storm takes much longer as it is well known from many news stories.

Building new transformers and substations (they are strongly connected) will also allow more power to be transported, thus allowing more distributed generation and energy-efficient equipment to be connected. Building new transformers reduces the amount of load per transformer, thus reducing the risk of overload. It also allows for shorter feeders, which, in turn, reduces the risk of overvoltages and undervoltages.

Building new primary infrastructure however gets more difficult. Next to the cost factor, there is the need to get permission from the relevant authorities. This holds especially for overhead lines where the public objection can be huge. The lead times can be long and once a line or cable is built there is very limited flexibility.

There are a number of classical solutions that do not require new lines, feeders or transformers. Solutions to mitigate voltage drops and rises include shunt and series compensation along the feeder, line-drop compensation on the transformer tap-changer, and automatic tap-changers for distribution transformers. The latter is not a classical solution, but the technology for it exists. With the exception of line-drop compensation, the costs could be rather high for these solutions.

Classical solutions to solve the problems with protection coordination, due to distributed generation, are directional protection and anti-islanding protection. Directional protection requires a voltage transformer, which increases the costs of protection substantially. Demonstration projects have also started where distance protection is used at a distribution level.

Anti-islanding protection is a cheap solution, but with large amounts of distributed generation, the risks of unnecessary trips could become too high. During two large disturbances in Europe (in 2003 and 2006), large numbers of distributed generation tripped the moment the system frequency dropped below 49 Hz, exactly according to the setting of the anti-islanding protection. In one case, the consequence was that about 15 million households experienced an interruption, about twice as much as when the distributed generation had not tripped. In several areas with large amounts of roof-top solar power, the anti-islanding protection trips on overvoltages on sunny days.

The harmonic emission from new types of production and consumption is often mentioned as a serious challenge. In fact, almost all types of new production and consumption are mentioned in this context: solar panels, wind power, small combined-heat-and-power ("micro CHP"), energy-saving lamps, adjustable-speed drives, heat pumps, and charging of electric cars. The classical solution for preventing high levels of harmonic distortion in the grid has been to set limits on the emission by individual equipment and by complete installations. For the low-voltage network, emission limits on individual equipment are used; for higher voltage levels, the limits on complete installations limit the distortion levels. Emission limits for individual equipment are set by the electromagnetic compatibility standards written by the International Electrotechnical Commission. Harmonics emission of small equipment, especially IEC 61000-3-2 [IEC, 2009], is of importance. For complete installations, no international standards exists; instead, it is the network operator that sets limits for the emission from individual installations, typically based on planning levels. These planning levels can also be chosen by the network operator, but, in practice, the indicative values in technical report IEC 61000-3-6 [IEC, 2008] are often used.

3.2 TRANSPORT CAPACITY

The availability of renewable energy is not equally spread throughout the world; even within a country or a continent, the distribution is far from homogeneous. Consider, for example, the European continent: hydropower is available in large amounts in the central parts (around the Alps), in Scandinavia, in northern Spain and on the Balkan. Wind power is available mainly in the west, especially the north-west, and solar power is most abundant in the southern parts of Europe. The heaviest concentration of consumption is found within the population centers in central Europe, in southern England, northern Italy, and with some of the national capitals. This is illustrated schematically in Figure 3.1. Transporting the production from large-scale renewable sources to the consumption centers will require an increase in transport capacity in the European transmission system. Using the balancing capabilities of hydropower will require further increase in transport capacity.

Even in the regions with high consumption and high availability of renewable energy, large installations will not be built close to the consumption because there is simply a lack of space to build. This holds, for example, for wind power in the low countries (from Northern France, through Belgium and the Netherlands, to Northern Germany and Poland); there is plenty of wind available, but the large installations will be built far away from the cities. The same is the case with solar power and cities like Madrid and Athens. The large solar power installations will be built outside the city although the need for energy is biggest inside of the city.

This is a problem that is not only related to Europe, nor is it only related to renewable electricity production. The distances between the main cities and the places with abundant renewable energy are possibly even further separated in North America and East Asia than in Europe. Any new, future generations of nuclear or coal-fired power stations will be located further away from the cities than the existing ones. Coal-fired power stations in Europe will more likely be built near major harbors. Also, the need for large amounts of cooling water makes locations close to the sea better. The same holds for nuclear power stations as they will be also be located as far away from cities as reasonably possible. In North America and China, countries with large coal reserves, it makes sense to build any new coal-fired power stations close to the mining areas and transport clean electricity instead of dirty coal. Future power stations with carbon-dioxide sequestration will more conveniently be built close to places where the carbon-dioxide can be stored.

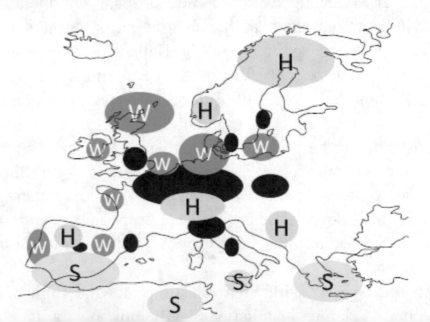

Figure 3.1: Distribution through Europe of consumption and renewable sources of energy. Black: consumption centers; grey: hydropower; green: wind power; yellow: solar power.

Finally, the creating of a pan-European electricity market requires the ability to transport large amounts of cheap electricity from one side of the continent to another. According to the vision of a European electricity market, customers should be able to buy the cheapest electricity wherever on the continent that it is located. The same developments are taking place in North America.

All this calls for the ability to transport more electrical energy over longer distances. The classical solution for this would again be to build more transmission lines, and in some cases, this is still the only solution. When a large hydropower station is built 2000 km away from where the electricity is needed, it cannot be avoided to build transmission lines. But there are methods of limiting the number of lines to be built or to increase the transport capacity of the existing transmission system without having to build new lines. This section discusses some of those methods.

The European TradeWind study looked at increased transport capacity between European countries towards the integration of 300 000MW of wind power into the European power system. This amounts to about 25% of the electricity demand. It was shown that increased transport capacity would reduce the operational costs by 1.5 billion Euro per year, justifying about 20 billion Euro in investment costs [Corbus et al., 2009]. It was shown in the same study that the capacity value of wind power increased from 8% for individual countries to 14% for the whole of Europe, assuming sufficient transmission capacity. That means that an additional 18,000MW (14 minus 8% of 300 000MW) of production capacity can be saved while keeping the same reliability. It also means that during peaks in consumption, there is more likely to be sufficient production capacity available, so that local price peaks occur less often. The result will be an overall reduction in the electricity price throughout the continent.

3.2.1 HIGH-VOLTAGE DIRECT CURRENT (HVDC)

An HVDC (high-voltage direct current) connection uses direct current (dc) instead of 50 or 60 Hz alternating current (ac) to transfer large amounts of electric power. When it comes to transferring large amounts of active power over long distances in a controlled way, there is nothing that beats the HVDC link. The power flow through the links can be fully controlled, completely independent of the flows elsewhere. The losses are mainly due to the converter stations and the additional losses per km are less than for an ac link with the same transport capacity. After a break-even point between one and a few hundred kilometers, HVDC is less expensive than an ac connection. Another important advantage of HVDC is that underground cables have no operational disadvantages, compared to overhead lines, as they do have for ac connection. This makes HVDC the only option for submarine connections. Also, for connections between two interconnected systems, HVDC is the only solution.

The main disadvantage of HVDC has been its high costs and the lack of experience of transmission-system operators with its integration in the ac system. The majority of HVDC lines in operation connect two synchronous systems. For example, the Nordic system is connected to the European system by means of eight links, with more links being discussed. Several links are also being discussed between Northern Africa and Southern Europe. An overview of the existing and some of the planned HVDC links in Europe is shown in Figure 3.2. The information for the map has been obtained from the ten-year network development plan as published by the organization for European transmission system operators (ENTSO-E). The development plan lists a total of 23 different HVDC projects to be commissioned within the next ten years.

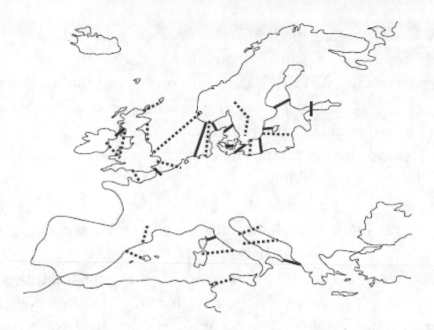

Figure 3.2: Existing (solid lines) and planned (dashed lines) HVDC links in Europe.

Looking at the map shows that the majority of the links is situated around the North Sea where the integration of wind and hydropower as well as further market opening are the main driving forces. There are also a number of links east and west of Italy. The lack of production capacity in Italy is an important driving force for these links.

In North America, HVDC links are mainly used as back-to-back links to connect the four big interconnected systems. There are also a number of long-distance links in Canada to transport hydropower from the North to the consumption centers in the South. In Japan, several HVDC links connect the two interconnected systems and, in China, several HVDC links are in operation to transport hydropower to the big cities.

HVDC links embedded in an ac system are less common. The 3100-MW Pacific intertie between Oregon and Los Angeles is a classic example. The two examples in Europe are the link between Sweden and Finland and the link between Italy and Greece. Both provide a short-cut through the sea between locations that are weakly linked through the ac grid. A fully embedded three-terminal link is planned between the South of Sweden, the middle of Sweden and the South of Norway. Two-terminal links are planned between Belgium and Germany, between France and Spain, and between Scotland and England.

A new development in the use of HVDC is so-called "VSC-HVDC" (VSC stands for Voltage Source Converter). There are several advantages with this modern type of HVDC, the main one being that the converters can be used to control reactive-power flow in the ac grid. The use of VSC-HVDC not only provides an efficient transport part in the form of the HVDC link, it also further increases the transport capacity of the rest of the ac grid in which it is embedded.

Future HVDC links are expected between countries and within countries, to transport large amounts of power between parts of an ac transmission grid but without having to strengthen the ac grid itself. The connection between Spain and France and the three-terminal link in Sweden are the first examples of this. The technology for HVDC and VSC-HVDC, is to a large extent, available. The technology for Thyristor-based HVDC is described in the classical work by Kimbark [1971] and by Arrillaga [1998]. The modern VSC-based technology is described by Arrillaga et al. [2007] and by Padiyar [2011]. Some applications of HVDC to improve the system stability are also discussed by Arrillaga [1998]. The future's main challenge will be in the dimensioning and operation of dc links embedded in an ac grid. The same holds for the FACTS controllers to be discussed in the next section. This is where new developments are expected to be.

The use of HVDC technology only partly solves the problem with the long lead times needed to build overhead transmission lines. Using dc, the additional costs of using cables instead of overhead lines are lower than with ac. The main difference is in the absence of reactive-power flows in a dc system. Cables however, also with dc, remain more expensive than overhead lines. Also, the much longer repair time of underground cables compared to overhead lines could be an issue. Conversion of ac lines to dc is a way to avoid the long lead times. Many of the same components can be used for a dc line as for an ac line; by adding a converter stations on each end, the line is able to transport 2 to 3 times as much power as before, without any of the stability issues associated with ac transmission. The barrier against this appears to be again the lack of experience in operating ac systems with embedded dc lines.

3.2.2 FACTS

One application of power-electronics converters, HVDC, was discussed in the previous section. The same kind of technology can also be used to control the flow of active and reactive power in an ac transmission system. The term "Flexible AC Transmission Systems" or FACTS is used for this. There is a whole spectrum of FACTS devices; for a detailed discussion on the technology and its applications, the reader is referred to Acha et al. [2002], Hingorani and Gyugyi [2000], Song and Johns [1999]. Summarizing, these devices can be split into two types: shunt devices and series devices.

Shunt devices control the reactive-power flow and have an important role to play in the voltage control. For example, by controlling the reactive-power flow over a transmission line, the risk of voltage instability can be greatly reduced. Alternatively, for the same risk, more power can be transported. The majority of FACTS devices currently in use are actually shunt devices, where the SVC (Static Var Compensator) forms the main part. Before the availability of large power-electronic converters, the synchronous condenser was used to control reactive power.

A series device changes the impedance of a transmission line or inserts a series voltage; in that way, the active-power flow is impacted. By changing the active-power flow, the transport capacity of the transmission system is not directly increased. It is however possible to reduce the flow through weaker parts of the grid. This in turn increases the overall amount of power that can be transported. In most cases, a phase-shifting transformer (or, "quadrature booster") is used for this, but several power-electronic solutions are under development. Phase-shifting transformers are also in use in Great Britain to steer the power flow from the north to the south.

Power-electronic converters, series as well as shunt and even HVDC converters, can also be used to damp inter-area oscillations and to prevent angular instability. Several applications of this are under development and part of ongoing research. Like with HVDC, mentioned in the previous chapter, the technology is to a large extent available, ongoing research and development is in the applications and the integration in transmission-system design and operation.

3.2.3 DYNAMIC LINE RATING

The maximum amount of power that can be transported over a transmission line can be set by a number of phenomena. For long connections at the highest voltage levels, it is often the risk of instability that sets the limit. But for shorter connections and for somewhat lower voltage levels the transport capacity is set by the maximum conductor or insulator temperature.

For cables and transformers, it is the maximum-permissible insulator temperature that sets the transport capacity. When the temperature gets too high, the ageing of the insulator will accelerate fast, resulting in a reduced lifetime of the component. To prevent this, the component is disconnected, manually or automatically, once the current exceeds a certain value for a certain time. This tripping current and time are chosen such that it is unlikely that the maximum-permissible temperature is exceeded for the local climate.

For an overhead line, it is the conductor temperature that matters. When the conductor heats up, it expands and the sag increases. The main consequence is that the clearance between the conductor and the surface, or between the conductor and objects, get too small. This can be a safety issue for persons passing under the line; but too close contact to a tall object like a tree can also result in a flashover, followed by a disconnection of the line. Also here, this is prevented by disconnecting the line once the current becomes too high for too long. The current setting is such that even for extreme weather situations (high temperature, low wind speed and high levels of solar radiation (insolation)), the conductor temperature does not exceed its maximum-permitted value. Some network operators use different current settings for summer and winter, but apart from that, the actual weather conditions are not considered.

With "dynamic line rating", the maximum current is adapted to the weather circumstances. On a cold windy day, the line will be able to transport more power than on a hot wind-still day. Most of the time, the line can transport more than the "static limit", but in rare circumstances, it can actually transport less. In this way, the line can be used to a much higher extent than using static limits.

The basic principle of dynamic line rating is rather simple; there is, however, some further development needed before it can be applied at a large scale. The main applications that are currently being studied are related to wind-power integration at subtransmission level [Etherden and Bollen, 2011, Kazerooni et al., 2011]. This is where the thermal capacity often sets the limit to the amount of wind power that can be integrated and where the biggest gain is expected. Two important remaining issues to be resolved are how to determine the actual limit, and who carries the risks.

As a first step, the existing method of using seasonal overload settings could be extended towards daily or even hourly settings, based on weather predictions and measurements. This would require certain margins, initially rather large margins, to accommodate for prediction errors and for variations in rating along the line. Especially, the wind speed (an important contributing factor to the conductor temperature) can vary significantly between locations. This could make that the line as a whole appears to have a high rating, while the conductor temperature gets too high for the part of the line that is shielded from the wind.

A more direct method would be to continuously measure the temperature and/or the conductor tension at several locations along the line and to use these either as an indicator of overload or as a basis for calculating the thermal rating of the line, see for example Albizu et al. [2011]. Direct measurement of the sag or the clearing between the conductor and the surface, or an object, would be a subject for future research.

The second issue to be solved concerns the quantification and management of the risks associated with the use of dynamic line rating. There are two types of risks here: the risk that the line rating at a certain moment in time is insufficient to transport the amount of power that needs to be transported, and the risk that the thermal rating of the line is overestimated. The former risk can be addressed by using an appropriate market mechanism (see Chapter 5) or a curtailment scheme (Section 4.4). The second risk needs to be incorporated in a safety margin between the estimated transport capacity and the triggering level for any protection or curtailment scheme.

A related development is the use of so-called "high-temperature, low-sag conductors". By using alternative conducting materials, the expansion of the conductor with a rising temperature can be reduced. This allows for higher temperatures and thus higher currents before the clearing between the conductor and the surface gets too low. By replacing the conductors, the transport capacity of a line can be doubled and such conductors are already in place at several locations. When using such conductors, the classical methods for setting of the overload protection are applied, resulting in a fixed or seasonal-dependent rating. A future development may see the combination of these conductors with dynamic line rating.

3.2.4 RISK-BASED OPERATION

The extremely high reliability of large transmission systems is mainly due to the method used for operating that system. This method is best summarized by the "(N-1) criterion"; the loss of any single component should not result in an interruption for any of the network users. No network user should notice that a major transmission line is tripped by its protection after a short-circuit fault. The same holds for the loss of a large production unit. This is called "secure operation". Should the system, after the loss of a large component, no longer be secure, the network operator takes measures to get the system back to a secure operation within a certain time, for example, 15 minutes or half an hour. The underlying principle is *"for a system to be reliable it should be secure most of the time"* [Kundur, 1994].

The (N-1) criterion has been functioning very well as can be concluded from the high reliability of most large transmission systems. The last major event in the European transmission system was in 2006; the last one in North America was in 2003. There are, however, two reasons for moving beyond the (N-1) criterion in some cases:

- During periods with extreme weather, single redundancy is not enough.

- During periods with mild weather, single redundancy could put an unnecessary limit on the transport capacity of the transmission network.

What fundamentally matters is not the weather but the failure rate of individual components. When the failure rate of individual components is sufficiently low, there is no need for any redundancy; when the component failure rate is high, single or even double redundancy is needed. This is schematically, and rather simplified, shown in Figure 3.3. For more details, the reader is referred to one of the books on reliability analysis for power systems [Billinton and Allan, 1996, Endrenyi, 1979] or for engineering systems in general [Barlow, 1998].

For a system without any redundancy, the system failure rate (the probability of a blackout in case of a power transmission system) is proportional to the component failure rate (the probability of loosing a transmission line, transformer, etc.): whenever a component fails, the system fails. With single redundancy, two components have to be out of operation at the same time before the system fails. When a component in the power system fails, repair will start soon after that. The system thus only fails when a second component fails before the first component has been repaired. The system failure rate is proportional to the square of the component failure rate and proportional to the repair time. The failure rate of transmission lines is normally very low, but during severe storms, the failure rate can increase by a factor of 1000 or more. The probability of losing two components within a short time could in fact become too high, so that additional redundancy is needed beyond the (N-1) criterion.

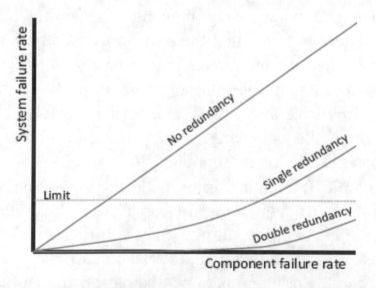

Figure 3.3: Relation between component failure rate and system failure rate for different amounts of redundancy (simplified).

The (N-1) criterion is, in fact, a "static criterion"; it is the same all the time. But sometimes the criterion is not sufficient and other times it is too much. Instead of a fixed amount of redundancy, the amount of redundancy should vary with the component failure rate. For example, during a severe storm, more redundancy should be available; whereas during periods with mild weather and the need for high amount of transported power, parts of the system could be operated without any redundancy. This is indicated in Figure 3.3 by means of the horizontal dashed line. For low component failure rate ("good weather"), no redundancy is needed, and the system can be used to its full capacity; for somewhat higher component failure rate ("normal weather"), single redundancy is needed; for high component failure rate ("extreme weather"), double redundancy or more is needed.

Risk-based scheduling of primary reserve (as part of the power-frequency control) is used for the operation of the Central Chilean interconnected system. The total costs of maintaining the reserve plus the expected costs of interruption are minimized. The resulting amount of primary reserve can be less than the size of the biggest unit; the system would, in that case, not be "(N-1) secure"; the loss of the largest unit would result in load shedding by the underfrequency load shedding scheme.

With any dynamic operational scheme (like risk-based operation or dynamic line rating), it is important to manage the risk if the transport capacity is insufficient. The risks to be managed are similar to the ones that have to be managed when using dynamic line rating (as discussed in Section 3.2.3). There is a risk that the transport capacity has to be reduced to a level that is insufficient for all the transactions that result from the electricity market. In that case, these transactions have to be restricted, which will typically give an increase in electricity price, either at the day-ahead market or at the balancing market. In some cases, curtailment of specific customers is a more appropriate solution.

When dynamic line rating is used during operational planning, the consequences of insufficient transport capacity are that planned transfers are not possible. This could result in market splitting with higher electricity prices as a consequence. In that case, the risk is carried by all consumers. It could also result in the production of a wind park being curtailed, so that the risk is carried by that wind park. With the approach used in Chile, the risk is carried by the network users that are disconnected by the under-frequency load shedding scheme.

The actual risk with risk-based operation is the occurrence of a regional or large-scale blackout. It will never be possible to afterwards decide if the calculated risk level was correct or not. Even though the probability of something happening is small, it can still happen. This brings us to the main issue: what is an acceptable risk level? Many more studies are needed before this question can be answered. An essential basis is the gathering of statistics to the variation of component failure rates with time. Once these failure rates are known, the calculation of the system failure rate is straightforward [Bollen et al., 2008].

Combining risk-based operation with fast and automatic curtailment schemes could be a way of getting experience without being able to accurately determine the system failure rate at any moment in time. This obviously requires an appropriate selection of the production and/or consumption to be curtailed. We come back to this in Section 4.4.

3.3 LARGE TRANSMISSION NETWORKS

We mentioned earlier the main challenges for the transmission system: to be able to connect the centers of consumption with the centers of production over a large geographical area so as to produce an international electricity market based as much as possible on renewable sources of energy. One of the ways of addressing this challenge has been discussed already in the previous section: increasing the transport capacity of the existing transmission system. There are two more possible solutions being discussed at the moment: both involve a continent-wide transmission network at a higher voltage level than the existing transmission network. The difference between the two is in the technology used: direct current (HVDC) or alternating current (50 or 60 Hzac). There are advantages and disadvantages with both technologies. In Europe, the discussions are mainly about an HVDC network, whereas in North America and China, the discussions are more about ac networks with operating voltages of 800 to 1200kv. An important reason for this may be that the most abundant sources of renewable energy (hydropower in Scandinavia, wind power on the British

Isles and solar power in Northern Africa) require a sea crossing to reach the main consumption areas. In North America and China, this is not the case.

In China, the main transport needs are from the resources in the west (hydro, wind and solar) to the large population centers in the east. Plans have been proposed for three large power corridors ("North corridor", "Central corridor" and "South corridor") with a transport capacity of 20 000MW each. Those corridors would consist of a combination of ac and dc links. A number of large HVDC links has been further proposed to integrate the hydro resources in Siberia with the consumption centers in countries like China, Korea and Japan. Recently, a plan under the name "DeserTec Asia" is proposing an HVDC grid to link China, Japan and Korea through several Southeast Asian countries all the way to Australia.

We saw in Figure 3.2 how existing HVDC links are grouped around the North Sea and the Baltic. This has resulted some years ago in the proposal to build a "Baltic Ring" and more recently in the proposal to build a "North-Sea grid". The aim of the Baltic Ring was to integrate the electricity markets of the countries around the Baltic Sea. The aim of the North-Sea grid is to link wind power, hydropower and consumption in the countries around the North Sea. Different versions of this grid have been proposed by organizations as diverse as the European Commission, Statnett (the Norwegian transmission-system operator) and Greenpeace. Several other proposals have been made as well. All proposals use the existing and planned links around the North Sea as a basis and the proposals are all limited to submarine links, as shown in Figure 3.4.

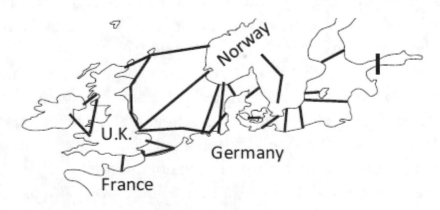

Figure 3.4: Vision for a North-Sea grid.

For a grid like this to be functional, it should be complimented either with a strengthening of the onshore transmission grid or the HVDC grid should be extended onshore so as to supply the electrical energy directly into the major consumption centers. Similar plans have been proposed for HVDC links across the Mediterranean to transport solar power from Northern Africa to Europe. Plans have also been proposed for a European-wide HVDC grid, covering the countries around the North Sea, the European mainland and Northern Africa. Although the idea is several years older, it only reached a large audience because of the presentation by the DeserTec consortium with plans to build large amounts of wind and solar power in Northern Africa and transport the electricity to Europe via an HVDC grid.

In Smith and Parsons [2007], reference is made to a "vision scenario" with 20% wind-power penetration in the United States. Part of this scenario is a 765-kV grid covering most of the country to transport wind power from the production areas to the consumption areas. The existing 765-kV grid in the Northeast of the country would be extended over the whole country. Back-to-back HVDC links would be in place at the borders between the three interconnected systems.

According to a study covering the eastern part of the United States, HVDC connections in combination with large ac "collector grids" are needed to transfer the energy from wind power in the Midwest to the cities on the east coast [Corbus et al., 2009]. In this scenario, over 200 000MW of wind power would be installed, more than half of this in the Midwest.

3.4 STORAGE

Peaks in production and consumption can be flattened by temporary storing electrical energy. By choosing the right location for such storage, the amount of renewable electricity production to be integrated in the local network can be increased. Storage can help in reducing peak load in general, but is especially discussed where it concerns integration of renewable energy and new consumption like electric cars. When adding a lot of electricity production as well as new consumption, peaks are expected to occur in both directions. Storage could become especially suitable in that case.

A hypothetical variation in power flow through a transformer is shown in Figure 3.5. The transformer supplies a combination of production and consumption. Both high production and high consumption can result in overload of the transformer. The transformer rating is given by the upper and lower dashed green lines. When the current through the transformer exceeds its rating due to a surplus of consumption, the storage should act as additional production, i.e., the storage should be discharged. The other way around: when it is surplus of production that causes the transformer to be overloaded, the storage should consume, i.e., charge.

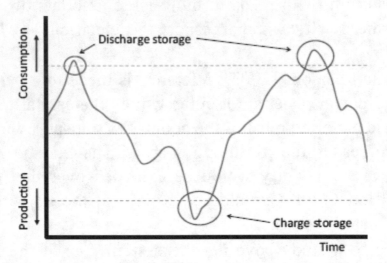

Figure 3.5: Variation in power flow through a transformer, including the need for storage.

When dimensioning and scheduling the storage, the following parameters are of importance:

- the amount of useful energy that can be stored; i.e., the difference between the maximum and minimum energy level (in kWh)

- the maximum charging rate (in kW)

- the maximum discharging rate (in kW)

- the losses during charging and discharging (in percent or in kW)

- the losses while the energy is stored ("leakage") including any energy needed for heating or cooling the storage facility (in kW)

Further factors affecting a decision about storage are the capital and operational costs as well as size, weight and life expectancy. An overview of the costs and other parameters for different storage technologies are summarized in Table 3.1 and Table 3.2. The data has been obtained from the Electricity Storage Association, from Walawalker [2011] and some other sources.

When comparing costs, the cheapest technologies are compressed air, electrochemical capacitors and pumped storage. The fact that the latter is still the one almost exclusively used is because those installations were built a long time ago. The operational costs of such installations are small.

Table 3.1: Comparison of the capital costs for some storage technologies

Technology	Capital cost	
	$/kW	$/kWh
NaS	1000–2000	200–1000
Flow Battery	700–2500	150–600
Li-Ion	700–1500	800–3000
Ni-Cd	500–1500	800–1500
Flywheel	4000–10000	1000–3000
Electrochemical Capacitor	200–600	100–200
Compressed air	500–1000	30–100
Pumped Hydro	600–1500	50–150

When comparing technologies, using the other properties in Table 3.2, we see that the electrochemical capacitor even has the highest efficiency and a very long lifetime. The energy density is, however, low, and energy levels above 10 kWh are not practical with this technology. Capacitors are used for some power-quality applications but not for limiting active-power flows

Lithium-ion batteries have become very popular the last few years due to their high efficiency and high energy density. A lifetime of 4000 cycles makes them suitable for daily use (it would result in a lifetime of more than 10 years). Costs remain high, and sizes small, but the expectations for this technology are high. A number of large installations of NiCd and NaS batteries already exist.

An installation based on Ni-Cd batteries is operating in Fairbanks, Alaska and able to supply 26MW during 15 minutes or 40MW during 7 minutes. A 34-MW, 245-MWh installation for wind power stabilization is in operation in Northern Japan. Other large installations are a 12-MW, 120-MWh installation in the UK, a 1-MW, 7.2-MWh installation using NaS batteries by American Electric Power (AEP) and three installations of 2-MW, 14.4-MWh at three different locations, also by AEP.

Table 3.2: Comparison of other properties for some storage technologies

Technology	Efficiency	Life Cycles	Density kWh/m^3	Rating
NaS	87%	2000	200	10 MW, 10 hrs
Flow Battery	80%	2000	25	1 MW, 6 hrs
Li-Ion	95%	4000	300	1 MW, 15 min
Ni-Cd	60-70%	1500	50	5 MW, 10 min
Flywheel	93%	20000	15	1 MW, 15 min
Electrochemical Capacitor	97%	30000	20	1 MW, 5 sec
Compressed air	75%	10000	-	100 MW, 10 hrs
Pumped Hydro	70-85%	20000	-	1000 MW, 24 hrs

An integrated utility, that has access to both the network and production could choose to invest in pumped storage or compressed air energy storage for transmission applications; however, for most network operators, those are no longer options. The various types of battery storage would be more appropriate for distribution level applications. A discussion has also started in some European countries on whether network operators are allowed to own and operate a storage installation. In Europe, network operators are not allowed to own production units. As a storage installation will function as a production unit part of the time, some have drawn the conclusion that the network operator therefore is not allowed to own a storage installation. An alternative reasoning is that a storage installation can make the grid operation more efficient, just like a capacitor bank for reactive power. That would result in the conclusion that network operators are allowed to own and operate a storage installation.

A related discussion concerns the further applications of the storage installation. Once the installation is available, a network operator could use the storage to charge the battery during low-price hours. That energy could be used to reduce peak consumption but also to cover the losses in a distribution network. The network operator might even go a step further by selling the energy in the storage during periods with high electricity price. To which extent, all this is allowed under the regulation in different countries is a matter of ongoing discussion.

3.5 ACTIVE DISTRIBUTION NETWORKS

In the previous sections, we have mainly discussed applications at transmission level. In this section, we will discuss a number of new applications at distribution level. The technological developments behind these applications are telecommunication and automatic control. The result is that new ways of protection, controlling or operating the distribution grid become technically and economically possible.

3.5.1 DISTRIBUTION-SYSTEM PROTECTION

The protection of the distribution network is impacted by distributed generation in a number of ways (see [Bollen and Hassan, 2011, page 299–366] and the references in there, for details):

- The short-circuit contribution from generators connected to a distribution feeder can result in an unwanted trip of the protection.

- The generator can also result in a reduction in short-circuit current, causing a fail-to-trip situation.

- The protection of the generator itself can fail to detect the fault, with an uncontrolled island operation as a result.

The first two impacts result in an increase in the number of supply interruptions for other network users. The last one is a safety issue, it could result in injury of maintenance personnel and damage to equipment.

The existing protection of the distribution system is based on local measurements only. A so-called "overcurrent relay" measures the magnitude of the current and compares this with a time-current curve. When the current exceeds a certain value for too long, the relay sends a signal to the local circuit breaker to open. This clears the fault, as well as interrupting the supply to all network users downstream of the circuit breaker. The threshold settings (current and time values) are determined during a "protection-coordination study" for all relays in the distribution network. Only when major changes are made in the distribution network are these settings updated.

Several more advanced methods for protection have been proposed and research is ongoing towards others. The number of demonstration projects and actual applications in the grid remains small. Research and development towards better protection methods is ongoing along the following lines:

- Protection based on local measurements only, without any communication with other locations. New types of directional protection are under development where advanced signal-processing tools are used to detect the direction to a fault. Using directional protection might solve the first two above-mentioned protection problems due to distributed generation. Some demonstration projects have started where the use of distance protection in medium-voltage distribution is studied.

- A method called "adaptive protection" has been proposed already around 1990 [Thorp et al., 1992], but only with more recent developments in communication technology has this become feasible on a large scale. The protection makes a decision (whether to trip the local breaker or not) based on local measurements. The threshold settings are, however, calculated centrally based on power flows and operational information (status of switches and breakers; presence of generators, etc.) and updated regularly. Any required changes are communicated to the individual relays. Instead of doing a protection-coordination study every few years, such a study is done every hour or even more often. A communication infrastructure is needed from a central processor (which could be in a control room or in an HV/MV substation) to all protection relays and between the central processor and the SCADA system. The communication is, however, not time critical. In case of failure of the communication, the relays could fall back to default settings.

- Instead of a central processor, individual relays could have a limited amount of communication with each other in the form of blocking, permitting, and intertrip signals. For example: when a relay at the start of a feeder sends a trip signal to the local breaker, it also sends a trip signal to all generator units connected to that feeder, so as to prevent island operation of the feeder. At the same time, this relay sends a blocking signal to the relays protecting other feeders from the same bus.

- Relays make decisions based on measurements at multiple locations. Differential protection, for example, is commonly used for important busbars and in some countries also for transmission lines. Differential protection requires a continuous exchange of data between the measurement locations. Another disadvantage is that it requires a separate back-up protection, whereas overcurrent-time protection has a "built-in" back-up protection. Some more advanced schemes that are being studied include the back-up function but still require continuous data exchange between locations.

- A compromise being discussed for transmission-system protection for many years now is to let the local protection take care of the most severe faults, using local information only and to use a more global scheme, using information from many locations, to take care of the milder faults and to provide the back-up function. Similar schemes could be developed for distribution networks with distributed generation. The setting of the "primary protection" based on local measurements would be such that all severe faults are cleared fast but at the same time that the probability of unwanted operation is low. The "secondary protection" based on global measurements would have more time to make a decision as the severe faults are cleared by the primary protection already. The settings of the secondary protection would be such that the probability of fail-to-trip is sufficiently low. Such a scheme could be combined with adaptive protection where the settings of both the primary and the secondary protection are recalculated regularly.

Next to the above-mentioned developments on protection of distribution networks with generation, there is much more research and development going on after the protection of transmission networks. A lot of the development is directed towards making protection faster and more reliable. When angular stability is the main limitation in the transport capacity over a transmission line or corridor, increasing the speed of the protection will increase the secure transport capacity. More reliable protection will contribute to an overall increase of the reliability of all network users.

3.5.2 VOLTAGE CONTROL

The voltage in a distribution network is maintained within acceptable limits, typically by using the following methods:

- An automatic ("on-load") tap changer on the HV/MV transformer keeps the voltage on the secondary side of that transformer within the dead band of the tap-changer controller. Some network operators use so-called "line-drop compensation" to adjust the position of the dead band to the loading of the feeder. This allows for more load to be connected and/or for longer feeders. The basic control principles remain rather similar.

- Some distribution transformers have a turns ratio that deviates from the ratio in nominal voltage between MV and LV. This results in a boost of the voltage at LV. For example, a transformer can have turns ratio 10.5kv/400V whereas the nominal voltages are 10kv and 400V. The result is that the voltage at LV, as a percentage of the nominal voltage, is 5% higher than at MV. This is also known as "off-load tap-changers". Some network operators use a series booster (for example, a 1:1.05 transformer) instead.

- The length of cables and overhead lines is limited; this holds at MV as well as at LV.

The principle is shown schematically in Figure 3.6. The solid red lines indicate the voltage drop along the MV feeder during minimum and maximum load. The green dashed lines indicate the range in voltage magnitude for two customers (A and B) connected to the LV network.

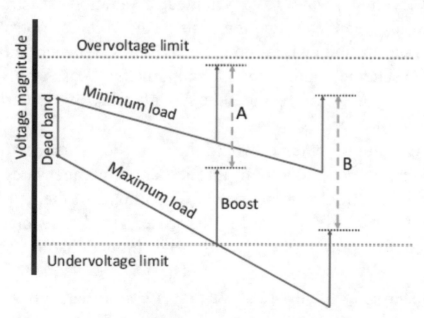

Figure 3.6: Principle of voltage control in distribution networks.

The setting of the on-load and off-load tap-changers and the feeder length should be such that the voltage magnitude always remains within a band bordered by an "undervoltage limit" and an "overvoltage limit". When the voltage comes outside of this range for too long, this is unacceptable.

This criterion should be fulfilled for all customers during maximum load as well as during minimum load. Adding production to the feeder will reduce the minimum load (and could even make the minimum load negative). In the figure this translates as a tilting of the minimum-load line upwards. For customer A already a small amount of distributed generation will result in a voltage above the overvoltage limit. In the same way, an increase in consumption will tilt the maximum-load line downwards, causing the voltage to drop below the undervoltage limit for customer B.

In both cases, the classical solution (as discussed in Section 3.1) is to strengthen or shorten the feeder. This remains the solution used in practice, but several alternative methods are being studied as part of ongoing research and development. A brief summary of some of those alternative methods is given below. Methods involving market principles and curtailment are discussed in Chapter 4. In this section, we will limit ourselves to schemes that do not impact the energy production.

For overvoltages due to distributed generation, a straightforward solution is to involve the production units in the voltage control. This is, in theory, possible for units with a power-electronics or synchronous-machine interface. The schemes studied involve, in almost all cases, only the control of the reactive power, but some schemes also involve control of active power during short periods.

A rather simple scheme would be for the generator to consume reactive power of such an amount that it exactly compensates the voltage rise due to the injection of active power. The impact of the generator on the voltage variations would be small in that case. The disadvantage would be additional reactive power flows, resulting in higher loading and losses in the distribution network.

A next step is to use the reactive-power control capabilities to directly control the voltage magnitude at the location where the generator is connected to the distribution feeder. Several schemes for this are being studied and proposed in the literature. With all schemes, the coordination between the voltage control by the generator and other automatic voltage control plays an essential role. The tap-changer is a rather slow controller with time constants of several seconds to minutes. When distributed generators react faster than the tap changer, they could end up injecting a high amount of reactive power and exporting this to the transmission system. To prevent this, the generator control should be slower than the tap changer, or some kind of coordination is needed.

The most common method used for coordinated voltage control is to have a central controller calculate the optimal voltage settings of the different control devices. Such schemes are also known as "volt-var control". This includes the automatic tap changers, capacitor banks, distributed generators, and any other converters involved in the voltage control. The optimization is such that the losses are minimized within the boundary conditions set by the overvoltage and undervoltage limits and by loading limits of lines, cables and transformers. Such schemes would be heavily based on communication: to get information on existing voltage levels at the point of connection with the network users and to communicate the voltage setting to the individual controllers.

In some of the schemes, the operating voltage is kept towards the lower side within the acceptable range. This will reduce the power consumption of several types of equipment and as such contribute to energy efficiency. This however does not work for heating or cooling load and for certain modern types of lighting. For those loads, the consumption is rather independent on the voltage, but the current and thereby the losses will increase for a lower voltage magnitude. Studies and demonstration projects performed in North America have shown that coordinated voltage control results for most feeders in a reduction of consumption up to a few percent.

A scheme that requires less communication uses a droop line where the reactive power production or consumption is based on the local voltage. If the voltage is higher than a given set point, reactive power is consumed; if the voltage is lower than the set point, reactive power is produced. Such a scheme could be combined with a rather large dead band in which no control of the voltage is needed.

3.5.3 MICROGRIDS

The microgrid is often presented as a way of enabling the integration of distributed generation in the electricity network and market. One of the original thoughts was to create additional economic advantages from the use of distributed generation. This would remove some of the economic barriers against distributed generation. The two main applications envisaged were improving the reliability and participating in electricity markets. Some of the original reasoning and the early results are presented in Deuse and Bourgain [2009].

Different authors use different definitions of the term microgrids. According to Chowdhury et al. [2009], a microgrid is a small-scale low-voltage supply network with CHP as the main production source and supplying electricity to one or more network users. The definition given by Hatziargyriou [2008] is more broader: microgrids are "low-voltage distribution systems with distributed energy resources such as microturbines, fuel cells, photovoltaic arrays, etc., together with energy storage devices and controllable load, offering considerable control capabilities over the network operation". According to Kroposki et al. [2008], a microgrid contains "at least one distributed energy resource and associated loads and can form islands in the electrical distribution system". The limitation to "low-voltage networks", does not seem to be necessary and several of the ongoing studies and recent experiments involve in fact medium-voltage networks.

Driesen and Katiraei [2008] distinguish between three types of microgrids: "utility-size microgrids", "industrial and commercial microgrids" and "remote microgrids". Remote microgrids are geographically remote and do not have any connection to the rest of the grid. This could be the grid on an island but also a very remote location. Such systems have been around for many years; they typically contain one or more flexible generators using fossil fuel (often diesel) for power-frequency control. Adding renewable energy and storage will reduce the diesel consumption. This not only reduces carbon-dioxide emission (and other emissions, too), it also often results in a significant saving because of the high costs for transporting the fuel to remote locations.

Industrial and commercial microgrids are located within the premises of one network user: this could be an industrial plant but also a commercial building like a university, a shopping center or a government agency building. These types of microgrids are discussed in Section 5.7.7. In this section, we will only discuss utility-size microgrids in which different network users contribute to the operation of the distribution grid. A utility-size microgrid is under the control of the network operator; the network users contribute to its operation, either voluntary or as a requirement under the connection agreement. The network users might also cooperate on the electricity market, forming a virtual power plant, but this is independent of whether they are part of the same utility-size microgrid or not.

A well-developed example of a microgrid (a rather big microgrid, in fact) is the cell-controller concept developed by Energinet, the Danish transmission-system operator [Martensen et al., 2011]. The cell controller keeps the reactive power flow between the local distribution network and the transmission grid constant. The set point of the controller is determined by the transmission-system operator. The control system is installed with the 150/60-kV transformer and communicates with all power plants downstream of the transformer (at 10 and 60kv). The control system receives information from the production plants and sends control orders back. A test of the control system was done in 2008 in a distribution grid with 10.8MW of CHP and 4MW of wind power [Lund, 2008]. A larger test was performed in November 2010 in a grid with 28 000 consumers, four CHP plants and 47 wind turbines. A final test was scheduled for July 2011. The results from these tests will be applied in the microgrid on the Danish island of Bornholm as part of an integrated European project. Several more microgrid experiments have been conducted as part of other European projects as well, including several utility-size microgrids.

An important property of all utility-size microgrids having been part of experiments is their ability to operate as a controlled electrical island independent from the main grid. Most of the research on microgrids is concentrating on the control issues related with that. The aim of controlled island operation is to improve the reliability of the supply to the network users. Next to the different technical challenges, for which we refer to the technical literature, there are some non-technical issues involved with this. One of them concerns who is responsible for continuity of supply, voltage quality and safety during island operation. Is it the network operator who is responsible or lies the responsibility with the network users that contribute to the island operation of the microgrid? Another regulatory issue concerns how to treat island operation and forced curtailment of consumption and production during island operation, in the continuity of supply statistics.

This becomes especially important when the income via tariffs is dependent on those statistics.

3.5.4 AUTOMATIC SUPPLY RESTORATION

When a fault occurs somewhere in the power system, the power-system protection will clear the fault by removing the faulted component. This is done by opening one or more circuit breakers. If the fault occurs in a radially-operated part of the power system, the opening of the circuit breaker will also result in all network users downstream of the faulted component experiencing an interruption. When the network is not only radial in operation, but also radial in design, the supply to those network users can only be restored once the faulted component is repaired. In fact, in many overhead distribution networks, one or more automatic restoration attempts are made. These "reclosing actions" are successful in 80% or more of the cases because many faults in overhead distribution networks are temporary and disappear after a few seconds [Bollen, 2000, pages 115–138].

Many distribution networks, especially those with underground cables and those at higher voltage levels and in areas with higher load density, are operated radially, but their design is meshed. This allows for restoration of the supply before the faulted component is repaired. This is especially important for underground cables, where locating and repairing a fault can take several days. The basic manual restoration principle can be explained with help of the simple network in Figure 3.7. The medium-voltage network shown in this figure consists of a number of radially-operated feeders originating from two HV/MV transformers (labeled "T1" and "T2"). Two feeders are shown in more detail in the figure; the end of these feeders are connected by means of a normally-open switch (labeled "N/O").

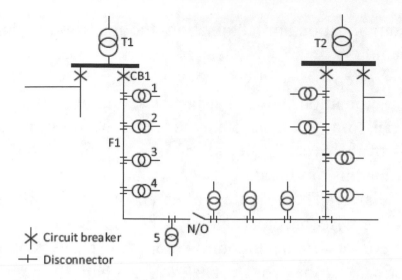

Figure 3.7: Radially-operated meshed distribution network.

Assume that a fault occurs at location "F1". The fault is detected by the protection with breaker "CB1" which opens and clears the fault. The result is that all network users supplied from distribution transformers 1 through 5 experience an interruption. With an overhead feeder, one or more reclosing attempts are made, for an underground cable not. When the fault remains after reclosing, or in case of an underground cable, a repair crew is sent out to locate the fault. Fault-current indicators are present with every one of the disconnectors on the medium-voltage feeder. In this case, the four indicators between circuit breaker "CB1" and the fault location will have indicated a fault current. In this way, it is found that the fault is between transformers 2 and 3. The disconnectors on both sides of the faulted section are opened, breaker "CB1" is closed and the normally open switch is closed too. In that way, all network users have their supply back before the repair of the faulted component has even started. Distribution transformers 1 and 2 are supplied from T1 as normal; transformers 3, 4 and 5 are supplied via a back-up route from T2.

All the readings and switching actions are done manually by maintenance personnel that have to come to the distribution transformers. In an urban network where the distances are relatively small, this can be done within 30 to 60 minutes, but in rural networks with long distances, locating the faulted section and doing the switching actions can take several hours.

By using communication and automation, the interruption duration can be reduced substantially.

The first step, typically referred to somewhat incorrectly as "distribution automation", is to establish communication between the distribution transformers and a control room. The opening of the breaker and the status of the overcurrent indicators is communicated to the control room. Based on this information, a crew can be sent out to the correct location immediately. The next step, often taken at the same time, is to use remote-controlled disconnectors and switches so that even the restoration process can be done from the control room. In this way, the duration of most interruptions can be brought back to less than 30 minutes. This may appear a long time for what are a number of simple switching operations, but the structure of actual distribution networks is much more complicated than shown in Figure 3.7. It is also often needed to wait for any recovery inrush to decay before the next switching action can be made.

A further step, the subject of recent research and development, is to completely automate the restoration process. The information from the circuit breaker and the overcurrent indicators will not be communicated to a control room but to an automatic system that makes a decision about which switching actions to perform. With an automatic restoration system, the duration of an interruption is expected to be reduced to just a few minutes. When the duration of the interruption is less than 3 to 5 minutes (depending on the local regulation), the interruption is no longer counted as part of the normal continuity of supply statistics. Instead, it might be counted as a short interruption, but those rarely have financial consequences for the network operator. This would give an important incentive for a network operator to invest in such an automatic restoration system.

3.6 MONITORING

An indirect, and very positive, consequence of much of the new technology being introduced in the electricity network is that measurement of voltage and current has become much more common than in the past, especially at lower voltage levels where measurements used to be extremely rare. Next to dedicated devices like power-quality monitors and disturbance power-electronic controllers and modern energy meters ("smart meters"). This availability of measurement data allows the development of various types of monitoring.

Monitoring applications can, in turn, help in addressing the various challenges. A number of examples will be described in this section. Some applications can be based completely on existing measurement equipment; for other applications, additional measurement devices may be needed.

So-called smart meters not only measure the cumulative amount of energy consumed, like the classical Ferraris meters, but also record the consumption over each shorter time interval (e.g., 15 minutes or one hour). This detailed consumption data can be used by a consumer to analyze its energy consumption. Peaks in consumption can be correlated with certain processes or activities. The next and most important step, after mapping the consumption, is to conclude from the consumption data how to reduce consumption. For domestic and office customers, expert rules or automatic learning techniques may be developed to simplify this process. For industrial installations, a manual process will likely be more effective.

The main driving force for the consumer will be to save money; the consequence for society will be a reduction in energy consumption.

The same data can be used to further save money when time-of-use prices and tariffs or peak-load tariffs are used. It might even be possible to reduce the subscribed power, which will often be a significant saving in electricity costs. The impact of all this on the network will be a reduction in peak load.

Consumption data with high time resolution is not only of use for the consumer, but this data also allows the network operator to make better investment decisions. Knowing the actual consumption patterns of individual consumers (including any local production) will give accurate information on the margins available for adding new production and consumption (the "hosting capacity"). There is no longer the need for keeping any reserves just to cope with the uncertainties in the existing consumption pattern. As a result, it is possible to connect more new production and more new consumption before investments in the distribution network are needed.

The same data will also reveal underestimation of the network loading. In that way, overloading can be prevented with a higher continuity of supply as a result.

Many smart meters can also record the voltage magnitude at the point of connection. This will give the network operator additional information on the hosting capacity for new production and consumption and on the need for additional measures when the voltage quality is insufficient.

Dedicated power-quality monitors take this a step further by not only recording voltage magnitude but also a range of other voltage and current disturbances. Additional information can be obtained from other measurement devices like protection devices and the before-mentioned smart meters, but with the current state-of-the-art, dedicated power-quality monitors are needed for most disturbances. Power-quality monitors can indicate when the voltage-quality becomes unacceptable and be used to estimate how much new production and consumption can be connected.

Disturbance recorders (also called "fault recorders" or "digital fault recorders") are used in most transmission networks for post-mortem analysis of faults and especially interruptions. The analysis of the recordings remains mostly a manual process. Automatic analysis of disturbance recordings will enable a much more efficient use of the data at all voltage levels. Such algorithms could also be used for the analysis of power-quality recordings and, for example, extract information on the cause of power-quality disturbances.

Automatic analysis of disturbances (either from power-quality instruments or from disturbance recorders) can also be used to detect deviations from expected behavior of the power system. Examples are a longer fault-clearing time than according to the protection coordination, larger angular oscillations after a fault than according to the simulations that were used to estimate the required operating reserve and less or more tap-changer operations than normal.

CHAPTER 4

Participation of Network Users

In this chapter, we will look at solutions that involve participation of the network users. This may be with compulsory involvement as part of the connection agreement or by curtailment where the network operator decides who to curtail. We will also discuss more voluntary curtailment schemes, but even with those, the network operator remains under control.

The first way of participation for network users is by the network operator to set compulsory requirements on production units; this is the subject of Section 4.1. We will next, in Section 4.2, discuss the circumstances under which curtailment of consumption already takes place in the existing system. In Section 4.3, a method called "intertrip" will be discussed, where a network user is disconnected immediately when a line is tripped. This is a rather coarse method of curtailment (we will refer to it later as "hard curtailment"), but it does result in improved utilization of the grid. Some of the emerging and future applications of curtailment and some of the related implementation issues will be discussed in Section 4.4. Finally, some numerical examples of the increase in renewable energy production by means of curtailment will be shown in Section 4.5.

4.1 SETTING REQUIREMENTS ON PRODUCTION UNITS

To be able to supply electrical energy to the consumers, the power systems need production units. The need of production units for supplying the active power is obvious, but the contribution of the production units is more than that. The production units play an important role in keeping the power system stable, among others by controlling reactive power flows. The control of the voltage is strongly related to this. Production units also ensure a sufficiently high short-circuit capacity, which in turn helps to limit voltage-quality variations and to ensure a sufficiently-fast operation of the protection in case of a short-circuit fault. Production units are also needed to ensure sufficient operating reserve so that the system remains stable even after the loss of a major production unit or transmission line. Finally, it is important that production units behave in a predictable way during major disturbances in the system, like short-circuit faults and the loss of a major production unit or transmission line.

All this can be taken care of easily when the electricity network (the "wires") and the production units belong to the same company, like was the case with the vertically integrated utility and is still the case with many industrial installations that supply their own electricity.

The separation of production, transmission and distribution changed the situation. Either the owner of the network or an independent system operator became responsible for maintaining reliability, voltage quality and security under a regulated monopoly, whereas the ownership of the production units was moved to an open market. This did not immediately result in any problems because the technical facilities for the production units to support the network were already in place; and even with new production units being built, this was seen as a natural part of the design and the cost of the units.

The new generation of production units is of a completely different nature from the large conventional thermal and hydropower units that made up the vast majority of production in the past. At distribution level, we find combined-heat-and-power units, solar power installations, and wind turbines. At higher voltage levels, wind parks and, in the future, solar parks will be connected. With such installations providing ancillary services to support, the grid is no longer a natural part of the design and merely seen as a cost. There are also a number of technical reasons that make it more difficult for these units to contribute, especially to operating reserves. Finally, the behavior of these new units during disturbances is often unknown. This holds for large wind parks, but also for the combined behavior of large numbers of small units.

To prevent the new production units from endangering the continuity of supply and voltage quality for the existing network owners, network and system operators set requirements on those units before they can be connected. Typically, the requirements are set by the operator of the network to which the unit is connected, thus the distribution-network operator in case of a small unit. Next to that, several system operators have started to set requirements on all production units, even those connected at lower voltage levels. The requirements from the distribution-network operator are mainly aimed at maintaining (local) continuity of supply and voltage quality as well as safety. Requirements differ strongly between network operators and between countries; in Europe, the reference document for very small units (up to 16 A) is European standard EN 50438 [CENELEC, 2007], whereas in the United States IEEE Std. 1547 [IEEE, 2003b] is the document to follow even for medium-size units. The requirements set in those documents include the following:

- Islanding detection to prevent the unit from powering a non-controlled island involving other network users. In practice, this results in setting of under/over voltage and under/over frequency protection so as to disconnect the unit from the grid whenever voltage or frequency deviates too much from its nominal value [Bollen and Hassan, 2011, page 342–358].

- Limits on the harmonic emission and contribution to voltage fluctuations from the unit. The aim of this is to prevent excessive harmonic voltages and voltage fluctuations for other network users. In EN 50438, reference is made to existing emission limits for (consuming) equipment.

- Limits on power factor. Most network operators prefer that small production units do not contribute to reactive power flow or voltage control. The most common requirement is therefore to maintain power factor close to unity.

- For larger units (above 250kvA in IEEE Std. 1547) monitoring of voltage, active and reactive power is required so as to keep the network operator informed of the status of the unit.

- Requirements on the immunity of the unit against voltage disturbances so as to prevent too frequent tripping of the unit.

For higher voltage levels, the requirements are part of a "grid code" or a "connection agreement" but also defined in separate documents issued by transmission system operators. Different requirements are in place in different countries, but a harmonization process has been started by among others ENTSO-E, the European organization of transmission system operators and ERGEG, the European organization of energy regulators [ENTSO-E, 2011, ERGEG, 2010b]. The following kind of requirements are in use in different countries and/or being discussed:

- Definition of the range of frequency and voltage parameters under which the production unit should remain operating as intended.

- Requirements for reactive power contribution to the grid. This could be a range of reactive power that is allowed but also following a specific request for reactive power from the system operator.

- Contribution of the production unit to load-frequency control including droop settings and requirements on spinning reserves.

- Contribution of the unit to short-circuit current. This is important for the stability of the system, for the correct operation of the protection, and for maintaining voltage quality.

- Requirements for the kind of protection devices to be present at the interface between the production unit and the grid and on the setting of these devices.

- Fault ride-through capability, so as to ensure that the production units remain available to support the system even after the loss of a large production unit or a major transmission line [Bollen and Hassan, 2011, page 437–447]. These requirements are also important for a predictable behavior of the production units after such a disturbance.

- Requirements on balancing capabilities and provision of ancillary services. These should typically be available on request from the system operator.

The underlying aim of the requirements at the distribution network is different from those at the transmission level, with as a result that the requirements sometimes contradict each other. At distribution level, the underlying philosophy has been that the production unit influences the network as little as possible. This explains, for example, the requirement that the unit should not contribute to reactive-power flow and voltage control and the requirement that the unit should disconnect fast when voltage and/or frequency deviate from their normal values. At transmission level, the underlying philosophy has been almost the opposite: the new production units should behave as much as possible like the conventional units. For example, at transmission level, it is compulsory for production units to contribute to reactive-power control. The reasoning, however, has been the same for both transmission and distribution: the presence of these new production units should not endanger the reliability and voltage quality for other network users, nor should it result in extra costs for these other network users.

Some of the grid codes and connection agreements distinguish between different types of production, for example, putting less demanding requirements on wind-power than on thermal and hydro units. In other documents, the requirements are only dependent on general properties like voltage levels and size of the unit. The latter is the prevailing trend because it allows for an equal treatment of all production units, independent from the energy source used. Here it should be kept in mind, however, that the requirements are often based on the capabilities of the existing large thermal and hydro units. For those units, the technology exists, and it is in most cases relatively easy to comply with the requirements.

For the new units, like large wind parks, the technology often does not exist and what might seem a reasonable requirement could become a barrier against the connection of a wind park. The current situation appears to be that the manufacturers of modern wind turbines have the situation under control and are able to comply with the requirements. Even the requirements on fault-ride-through, that seemed to be a serious problem at first, are to a large extent under control. In fact, instead of limiting the building of large wind parks, the requirements have triggered a lot of research and development on many different solutions to improve the fault-ride-through of individual wind turbines and complete wind parks.

The requirements on balancing and power-frequency control can be easily fulfilled with modern technology, but there is a serious drawback. To be able to participate in frequency control and balancing, a wind farm needs reserve capacity above its produced power. As the total capacity (production + reserve) is determined by the wind speed, the farm is not using all the energy from the wind that is available. This energy will have to be produced from other sources, typically fossil fuel. The participation of wind parks in frequency control and balancing will thus result in more fossil fuel to be used. Obviously, when the choice is between disconnecting the wind park or having it connected but below its maximum capacity, the latter is to be preferred. But the network operator, not willing to discriminate between generators, may share the reserve requirements over all units at all times.

At distribution level, the main challenge remains to cope with the requirement on islanding detection. The existing methods remain based on detecting abnormal voltage and frequency at the point of connection and network operators generally accept this as sufficient. However, also here the requirements have triggered a lot of research, but not much of this has reached the development stage yet.

4.2 CURTAILMENT: EXISTING APPLICATIONS

Curtailment of production or consumption is not a completely new subject. In fact, it has been a common tool for the system operator to prevent the system from collapsing. At production side, every system operator has the possibility to intervene in the electricity market when this would result in non-secure operation of the system. There are two further ways of curtailment of consumption in use to protect the transmission system against collapse. These are "rotating interruptions" and "underfrequency load shedding". The former is used when the operational security becomes insufficient, the latter as a last emergency measure.

Rotating interruptions are rather indiscriminate. Important installations are supposed to be equipped with back-up installations, but this is no longer countries that even those responsible for important installations like hospitals no longer expect the supply to fail. Recent development are towards not interrupting feeders containing important customers. In the future, it will be able to prioritize at the level of individual customers. In some countries, the system operator has agreements with certain consumers that will be curtailed first before starting rotating interruptions that involve all consumers. This could be developed further into a market for interruptable loads.

Rotating interruptions are a way of ensuring the operational security of the system during periods with a shortage of production and/or transport capacity into a certain region. When rotating interruptions are in place, there does not have to be an actual shortage of production capacity. Instead, the criterion is that there is not enough production capacity, above the maximum expected consumption, to provide operational reserves. Rotating interruptions can be due to high demand typically associated with extreme weather. Rotating interruptions were, for example, in place in Italy on 26 June 2003 when the country suffered from extremely high temperatures. In Texas, rotating interruptions were needed in February 2011 after the loss of a number of power stations due to extreme weather. Rotating interruptions are also needed when there is a shortage of production or of transport capacity. An example of the latter is the use of rotating interruptions in Manila, Philippines after a fire in a large transmission transformer in August 2009. Shortage of production, among others due to dry weather, was behind the rotating interruptions during the California electricity crises of 2000 and 2001.

Underfrequency load shedding, the second method for curtailment of consumption, is only used when all else fails: there is an actual shortage of production, and because of that, the frequency in the system will drop fast. To prevent a complete collapse of the system, under-frequency relays trip part of the consumption so as to bring the remaining consumption in balance with the production again. The amount of consumption tripped can be several tens of percent of the total consumption, as can be concluded from the example settings in Table 4.1 and Table 4.2. In the first example (used in Florida), load shedding starts already when the frequency deviates 0.3 Hz from its nominal value: when the frequency drop below 59.7 Hz, 9%

of the load is disconnected; when the frequency drops below 59.4 Hz, another 7% (for a total of 16%), etc. The different frequency limits are rather close to each other and when the frequency reaches 59 Hz, a total of 56% of the load is disconnected.

The second example of setting of under-frequency load shedding is the one commonly used in the European interconnected system where there are much larger steps between the frequency thresholds. Disconnection of load starts when the frequency drops 1 Hz below its nominal value. When the frequency has dropped 1.5 Hz, between 30 and 50% of load is disconnected.

Underfrequency load shedding is rarely activated in large interconnected systems. A recent example occurred in Australia on 2 July 2009 when the failure of a current transformer at an unfavorable location resulted in the loss of 3000MW of production. A total of 1130MW of load was disconnected by the under-frequency load shedding, all of which was reconnected in about one hour [AEMO, 2009]. On 4 November 2006, a total of 17,000MW of load was tripped by under-frequency load shedding in Western Europe after the overload of a number of lines in Germany resulted in system splitting with a large shortage of production in the Western part of the system [UCTE, 2006]. Also, in this case, the supply to all customers was restored in about one hour. Without under-frequency load shedding, both events would have resulted in a massive blackout for almost all customers and many hours, if not days, to restore the supply.

Table 4.1: Example of settings for under-frequency load shedding - small margin between steps

step	frequency	delay	amount of load
A	59.7 Hz	0.28 s	9%
B	59.4 Hz	0.28 s	7%
C	59.1 Hz	0.28 s	7%
D	58.8 Hz	0.28 s	6%
E	58.5 Hz	0.28 s	5%
F	58.2 Hz	0.28 s	7%
L	59.4 Hz	10.0 s	5%
M	59.7 Hz	12.0 s	5%
N	59.1 Hz	8.0 s	5%

Table 4.2: Example of settings for under-frequency load shedding - large margin between steps

frequency	amount of load
49.0 Hz	10-20%
48.7 Hz	10-15%
48.5 Hz	10-15%

At distribution level, curtailment is less often in place, but in principle, it can also be used here. One might imagine a damaged transformer that may require several months to repair. If a high load situation occurs during this period, the network operator will have to apply rotating interruptions (i.e., to curtail consumption) to prevent an interruption for a larger part of the customers. Also some network operators offer "curtailable tariffs": such a tariff is lower than the normal tariff in return for a reduction in load upon request by the network operator.

The Swedish and Finnish transmission system operators have access to a certain amount of reserve that is kept outside of the electricity market. This is partly in the form of production units that are standing still most of the time and partly in the form of guaranteed amounts of demand reduction. For the winter 2011/2012, the Swedish transmission system operator has access to 1726MW of reserve, 362MW of which is in the form of reduction in consumption. The latter is spread over four industrial installations and two aggregators. During the coming years, the volume of this reserve will be reduced and it will have to consist completely of demand reduction.

4.3 INTERTRIP

Intertrip is a method of shifting reserves from the network to the network users for certain events. Instead of building new lines, the operating reserve is obtained in the form of an immediate reduction of production or consumption once a component fails. Intertrip has been used for many years as a way of allowing more production to be connected to a location in the network than would be possible otherwise. The term "intertrip" refers to the transfer of a trip signal from one switch or circuit breaker to another; it is used, for example, to speed up the operation of protection. The term "pilot protection" is also used in this context [Anderson, 1999, Horowitz and Phadke, 1995]. Here we will only consider intertrip where it is used to disconnect network users (production or consumption) when a component in the grid fails or is removed. Instead of "intertrip", the terms "system integrity protection scheme" and "remedial action scheme" are in use [Bakken et al., 2011].

4.3.1 INDUSTRIAL INSTALLATION

The concept of intertrip is explained here for a double-circuit line supplying an industrial installation and a small town, as illustrated in Figure 4.1. We will first explain the principle for curtailing consumption using intertrip, although the method is more likely first to be used for curtailing production.

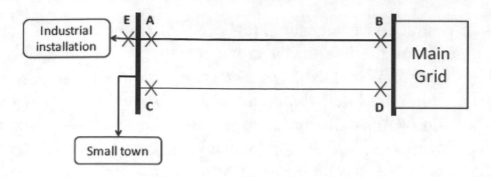

Figure 4.1: Double circuit line with two types of consumers.

The double circuit line in Figure 4.1 supplies power to a small town and to a rather large industrial installation. The classical design of the network would be such that even one of the two circuits is able to supply the maximum consumption of both the town and the industrial installation. The transport capacity of the line would be twice the total maximum consumption. This transport capacity would, however, only be used in the case that the outage of one circuit corresponds with the maximum consumption.

The alternative is to reduce the consumption, using an intertrip signal, whenever one of the circuits is out of operation. As a first step, we assume that the reduction in consumption takes place by disconnecting the industrial installation. With reference to Figure 4.1, this would mean for breaker E to be opened when one of the breakers A, B, C or D is opened. In practice, this is implemented by installing a contact on those breakers so that their opening automatically creates a (inter)trip signal for breaker E.

With such an intertrip scheme, there is no longer a need to consider the worst case in the design of the line. Instead, the following two design criteria are in place: each circuit should be able to supply the maximum consumption of the town; the two circuits together should be able to supply the maximum consumption of the town and the industry. If the maximum consumption of town and industry is about the same, the transport capacity of the line has to be only about half of what it would have needed to be in the case without a transfer scheme.

There could be a number of reasons for introducing such a scheme. When building a new line, the line is normally designed to cover the worst case. The main applications of intertrip schemes are when significant changes in production or consumption occur after the line has been built. For example, when the line has been built to supply the town only and the industrial installation is added later. The costs for strengthening the line, or building an additional line, would normally be borne by the new customer, i.e., by the industrial installation. It could also take several years to increase the transport capacity, which would delay the start of production in the industrial installation. An intertrip scheme gives a serious cost saving and speeds up of the connection to the grid.

There are some drawbacks with a scheme like this. The industrial customer will experience a lower reliability. Whenever one of the two circuits is out of operation, the industrial installation will be out of operation. The unavailability and interruption frequency as experienced by the industrial installation are thus twice as high as in the case when there would be just a single circuit. As industrial installations are known to require a high reliability, this may well be unacceptable in the long term. For some types of industrial installations, there are also safety issues associated with interruptions; in that case, an intertrip scheme like this would almost certainly not be acceptable.

But the good thing is that the costs of interruptions can be weighed against the costs of upgrading the line. Even the costs of additional safety or reliability measures can be taken into consideration. When more industrial customers are connected to the same substation, lower tariffs can be offered to those that are willing to accept a lower reliability. This offer might even be extended to domestic, commercial and small industrial customers in the town. In such a case, additional communication infrastructure is needed.

There are also disadvantages to this scheme from the viewpoint of the network operator, apart from the costs of the communication infrastructure. In case of failure of the intertrip signal, both circuits will be lost. This will result in an interruption for the town as well. The use of the intertrip scheme will thus also result in a reduced reliability for the customers not involved in the scheme. It is difficult to make an accurate estimation of how often this would occur, but there is certainly a risk involved here. The probability of this happening might not be very high, but it should be noted here that this involves a voltage level at which the (N-1) criterion is applied. That means that interruptions are actually not acceptable at this voltage level:

any interruption is one too many. As it is the network operator that is going to be blamed for an interruption due to failure of the intertrip scheme, many network operators are understandably somewhat reluctant to introduce such schemes. However, despite this reluctance, intertrip schemes are applied more often in the connection of wind parks, as we will discuss in the next section.

4.3.2 WIND PARK

The same intertrip scheme as shown above can also be used to allow more wind-power production to be connected to a subtransmission network. A simple example, again for a double-circuit line, is shown in Figure 4.2. Like in the previous case, circuit breaker E is opened automatically whenever one of the breakers A, B, C or D is opened. The result is that the wind park is disconnected whenever one of the circuits is not in operation.

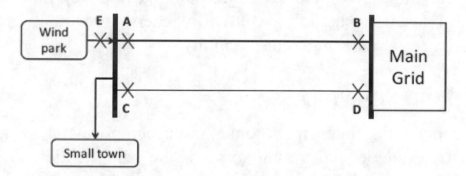

Figure 4.2: Double circuit line with consumption and production.

If we assume that the line has been built to supply the town, the transport capacity of each of the circuits will be at least the maximum consumption of the town. For a small amount of wind-power production, the loading of the line will become less because production and consumption compensate each other (in effect, the wind park supplies part of the town). But during periods with low load and strong wind, some of the wind power will flow back to the main grid. When this back flow becomes too big, the line will get overloaded.

The same design criterion is also here that the line should be able to supply all customers in all cases even when one circuit is out of operation. The transport capacity of each circuit should thus, at least, be the maximum production of the wind park minus the minimum consumption of the town. Or, when the line is considered as given (which is typically the case), the maximum production of the wind park should be less than the transport capacity of one circuit plus the minimum consumption of the town. This is called the "hosting capacity" of the grid for new production.

By using an intertrip scheme, the hosting capacity can be increased. The outage of one circuit does not have to be considered anymore because in that case the wind park is disconnected anyway. With two circuits in operation, the maximum production can be as high as twice the transport capacity of one circuit plus the minimum consumption of the town. The hosting capacity has thus been increased by the transport capacity of one circuit without having to actually build anything beyond some new communication links. Another advantage is that such an intertrip scheme can be installed rather quickly, a matter of months at most; whereas planning, obtaining permission for, and building of a new line may takes years. The main drawback of the intertrip scheme remains that the wind park experiences a rather high unavailability.

4.3.3 MESHED NETWORK

In meshed networks, intertrip becomes more complicated, but it remains possible. An example, illustrating some of the complexities, is shown in Figure 4.3 and Figure 4.4: the double-circuit line between A and B is a main transmission corridor, but part of the power also flows though substations C and D. As explained in Section 4.3.2 of Bollen and Hassan [2011], this can result in the somewhat counterintuitive situation that reducing consumption at substation C or D increases the flow between C and D. Thus, adding a wind park at location C would reduce the transfer capacity through the main transmission corridor. To prevent this, an intertrip scheme can be used that disconnects the wind park, as in Figure 4.3. In some cases, the problem can be solved by operating part of the network radially as shown in Figure 4.4. This however requires additional studies and, more importantly, will reduce the reliability for all network users connected to substations C and D.

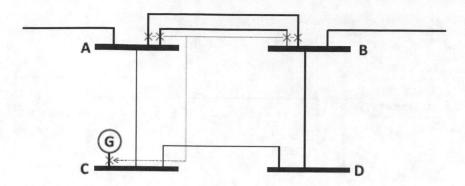

Figure 4.3: Intertrip scheme in a meshed system: tripping of production.

4.3.4 OVERLOAD-TRIGGERED DISCONNECTION.

As was mentioned before, an important disadvantage of the intertrip scheme is that the customer to be tripped experiences a reliability that is even less than for a radial feeder, at a voltage level where normally the (N-1) criterion is used. Let us return to the original need for reducing the load: the power through the line exceeding the transfer capacity of the line. Instead of simply tripping the customer during every circuit outage, disconnection is only needed in those situations in which there is actually a risk of overload. The principle of such a scheme is shown in Figure 4.5.

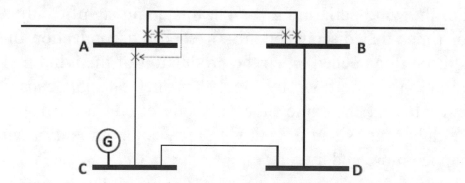

Figure 4.4: Alternative intertrip scheme in a meshed system: radial operation.

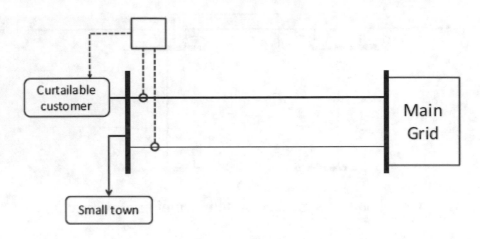

Figure 4.5: Overload-triggered disconnection for a double-circuit line

Upon detecting an overload, the protection normally trips the component that is to be protected: one of the circuits of the line. But that would result in further overloading of the parallel circuit. With overload-triggered disconnection, the source of the power flow through the circuit is disconnected instead. This prevents cascading and limits the interruption to one or more selected customers. The result is that the load is only disconnected when the circuit is actually overloaded. For an industrial installation with rather constant consumption, the gain would be rather limited. For the wind park, in Figure 4.2, this could significantly reduce the number of times that disconnection is needed. Disconnection of the wind park would only be needed when the production of the wind park exceeds the consumption of the town by more than the transport capacity of one circuit, and when at the same time only one circuit is in operation. This might be a rather rare situation, but with increasing amount of wind power, it could happen more and more.

An additional advantage of overload-triggered disconnection is that it also covers cases where, for example, the wind park produces so much power that even two circuits are not enough. But that would reduce the reliability as experienced by the wind park even more. It should also be noted that disconnection most likely takes place when the wind-power production is high. Thus, even if disconnection takes place only during a small percentage of time, it could still correspond to a rather high percentage of energy being spilled. We will show some numerical examples of this in Section 4.5.

4.4 MORE APPLICATIONS OF CURTAILMENT

4.4.1 PREVENTING OVERLOAD

The term "overload" should be seen in a wide sense here. It includes not only thermal overload but also situations when the voltage magnitude is out of its acceptable range, instability in the transmission system due to excessive power transfers, and insufficient operating reserve for secure operation of the system. Different methods for detecting the occurrence of an overload situation are needed for the different types of overload. For example, thermal overload can be detected by measuring the current or by measuring the temperature, whereas an unbalance between production and consumption in an interconnected system can be detected by measuring the frequency. The various methods that could be used for activating curtailment will be discussed for some of the applications below.

An important difference in detecting the overload is between curtailment that is only activated when the overload situation is actually near and curtailment that is already activated when there is insufficient operating reserve. This distinction is already present in the existing methods for curtailment: rotating interruptions are activated when there is insufficient operating reserve, whereas under-frequency load shedding is only activated when there is an actual overload.

In radial-operated networks there are no operating reserves and curtailment can only be activated when the overload is actually near. This type of curtailment can make use of an actual measurement of the loading of the component (e.g., the current through a cable in case of thermal overload). In many cases, there will be an actual loading limit that should not be exceeded because it could lead to serious equipment damage or dangerous situations. There will also be a protection setting that trips the component when its loading gets too close to the actual loading limit. In case of thermal overload, this is a combination of time and current settings: when the current exceeds a certain level for longer than a certain time, the component is tripped. (This is known as "inverse-time overcurrent protection".) Curtailment should be activated before the component will be tripped by its overload protection. An example of this, for thermal overload, will be shown in Figure 4.8 and Figure 4.9, below.

When curtailment is used to prevent the operating reserve from getting insufficient, the setup of the scheme is completely different. Such schemes can only be applied when the network is operated meshed as is common at higher voltage levels, transmission and subtransmission, where reliability is very important. Supply interruptions are all but unacceptable at these voltage levels; therefore, it is common to use a certain amount of operating reserve. To maintain secure operation, the curtailment scheme needs to keep track of the amount of operating reserve that is available and on the amount of operating reserve that is required. A possible curtailment scheme is summarized in Figure 4.6.

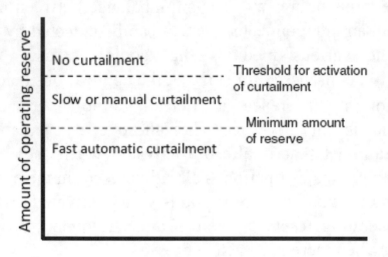

Figure 4.6: Thresholds for curtailment when aimed at maintaining sufficient operating reserve.

The minimum level of operating reserve needed is set in the rules or guidelines for the operation of the system, where in practice the operating reserve is typically chosen such that the loss of any component can be coped with. Once the amount of operating reserve is less than this minimum-acceptable value, the operator will intervene one way or the other to make more operating reserve available. Under the existing operational rules, a certain amount of time is available for this, where 15 minutes is a typical value. This is however based on the understanding that situations with insufficient operating reserve are rare. When the system is operated closer to its (secure) limit, violations of the minimum amount of reserve will occur more often and 15 minutes times to restore the reserve may no longer be sufficient. Instead, in the curtailment scheme, production or consumption is curtailed fast and automatic once the reserve becomes less than its minimum limit. Next to that, a threshold is set at a higher level of reserve; once this threshold is exceeded, curtailment is activated, but this may be slow or manual curtailment. The above-mentioned 15-minutes rule may still apply to the higher threshold.

The next decision to be taken by the curtailment scheme is to decide how much curtailment is needed. Here we will distinguish between "hard curtailment" where one or more network users are disconnected completely and "soft curtailment" where the curtailment is only the amount needed to keep the loading below an agreed limit. Hard curtailment is easy to implement; in fact, the intertrip schemes discussed in Section 4.3 are examples of hard curtailment. To implement soft curtailment, a calculation is needed to determine what this minimum amount of curtailment should be. This requires knowledge of the system so as to find a relation between the amount of curtailment and the reduction in loading. When curtailment is used to prevent overcurrent in a radial network, the amount of curtailment needed is easy to calculate. But for other parameters and in meshed networks, the calculation could be rather complicated. The development of methods for risk-based operation in combination with methods for fast estimation of system stability is needed before a good curtailment scheme can be implemented, see Section 3.2.4.

An alternative is to use a control algorithm as for example shown in Figure 4.7. An overload indicator obtained from measurements in the power system is compared with a curtailment activation limit. When the overload indicator exceeds the limit, a curtailment signal is communicated to a production unit. Controller "Co2" is a non-linear controller; its output is zero as long as the overload indicator is less than the curtailment activation limit (an agreed value at which curtailment is activated). The more and the longer the limit is activated the stronger the curtailment system. The tuning of both controllers ("Co1" and "Co2") should be such that the feedback loop is stable and that the ramp rates after "Co1" do not damage the production units, but also such that the loading of the component remains below an agreed overload curve (e.g., the setting of the overload protection).

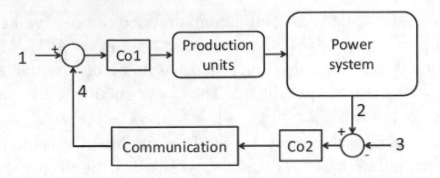

Figure 4.7: Control system for soft curtailment: 1 = production; 2 = overload indicator; 3 = curtailment activation limit; 4 = curtailment signal.

4.4.2 THERMAL OVERLOAD IN RADIAL NETWORKS

Thermal overload is the main limitation for transporting power in radial networks. With increasing amounts of new production and new consumption, it will be the thermal capacity of overhead lines, underground cables and transformers that determines how much can be connected before investments in primary infrastructure are needed. Postponing such investments or possibly not having to make them at all would be a significant saving. This of course assumes that there is a cost-effective alternative, e.g., in the form of curtailment. The steps to take when designing a curtailment scheme against thermal overload are the following ones:

- Choose the overload indicator.

- Agree on an overload activation limit.

- Decide what to curtail and how much.

Consider as an example a curtailment algorithm to prevent a transformer from getting overloaded. The curtailment algorithm will reduce consumption or production when the current through the transformer gets too high for too long. The setting of this algorithm should be such that the disruption (i.e., the amount of curtailment) is as small as possible, but at the same time, the curtailment should be fast enough so that the overload protection of the transformer does not remove the transformer from service.

This is a classical protection coordination problem, as shown schematically in Figure 4.8.

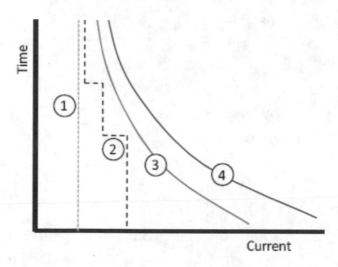

Figure 4.8: Curtailment settings in relation to protection settings. (1 = maximum permanent load current; 2 = curtailment settings; 3 = overcurrent protection; 4 = actual thermal limit).

Thermal overload of a transformer is a relatively slow process. For example, recommendations for the setting of short-circuit protection give a value of 2 times the rated current for 30 minutes [IEEE, 1993]. However, once the setting of the overload protection is known, curtailment has to be coordinated to this setting. For hard curtailment, exceeding the limits according to curve 2 in Figure 4.8 would result in one or more customers or equipment at customer installations to be tripped immediately. For soft curtailment, the amount of curtailment would be less.

Curtailment of production is a way of taking the risk due to production units away from the network operator and other network users. There is, however, a certain risk remaining, which has to do with the finite probability that technology does not function as intended. When, during a potential overload or overvoltage situation, the curtailment does not work as intended, the production will not be curtailed and other network users may suffer an interruption. Although the probability that this happens may be small, the consequences could be big, especially when using curtailment at transmission level. Fear for the consequences of failure of the new technology may also become a barrier against introducing this new technology. A possible solution to limit the consequences of such a failure is to add an additional layer of protection. Consider again overload of a transformer in a radial network as an example, for which the protection coordination is shown in Figure 4.8. When the curtailment would not work (curve 2), the next line of protection would be the overload protection (curve 3), resulting in tripping of the transformer and most likely an interruption for all network users downstream of the transformer. To prevent this from happening, an additional overcurrent relay can be installed to protect the transformer, shown in the blue dashed curve (labeled 5) in Figure 4.9. This relay would however not trip the transformer but one or more of the wind-power installations downstream of the transformer.

Installing this additional level of protection would require more margin between the curtailment criterion (curve 2) and the actual overload protection of the transformer (curve 3). Shifting curve 3 to the right would increase the risk for the network operator and for the other network users, which is what we were trying to prevent. The only solution, in case of insufficient margin, would thus be to shift the curtailment curve (curve 2) to the left. The consequence of this additional layer of production would thus not only be the probability of loss of production due to a curtailment failure, but also more overall curtailment. However, this may be needed, at least in the beginning, to get large-scale use of curtailment acceptable for the network operators.

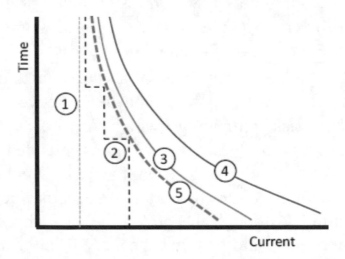

Figure 4.9: Additional level of protection (curve 5) to prevent transformer tripping due to failure of the curtailment.

4.4.3 THERMAL OVERLOAD IN MESHED NETWORKS

In a meshed network, the detection of overload is the same as in a radial network. Here the actual current, a dynamic current limit, the temperature or an estimation of the sag of the conductor could be used as an overload indicator. Once the overload is detected, the situation becomes completely different. A possible curtailment scheme could be as follows:

- The active and reactive power flows through all lines are continuously monitored. Also weather parameters are monitored so that the maximum-permissible current through each line is known.

- The impact of the outage of any single line on the power flows through the other lines is calculated.

- When any of these line outages results in the maximum-permissible current being exceeded for any of the other lines, curtailment will have to be activated upon the outage of that line. The need for curtailment of production and/or consumption is calculated.

- The information on required curtailment is communicated to the circuit breakers with that line. Upon opening of a circuit breaker with that line the local curtailment data is used to communicate curtailment signals to the relevant network users.

What is mentioned here for overhead lines holds also for underground cables and transformers. However, the impact of weather parameters is less so that the gain obtained by dynamic rating will be less. Once a possible overload is detected in the third step above, the network would no longer be secure if there were no curtailment in place. Curtailment is used here as a way of shifting the reserves from the network to the network users.

4.4.4 OVER- AND UNDERVOLTAGES IN RADIAL NETWORKS

In rural distribution networks, the limit to the transfer of power is not the thermal capacity of a feeder but the voltage drop. When connecting distributed generation, overvoltage could also occur due to the voltage rise.

The same three issues as before come up with a curtailment scheme: the choice of overload indicator, the overload activation limit and the amount of curtailment.

The actual voltage magnitude with the terminals of the network users is the best indicator. When the voltage with one of the network users is outside of acceptable limits, curtailment should be activated.

When overvoltages occur the easiest option is to accept a limited risk of equipment damage due to overvoltages and allow an increase on the amount of distributed generation. Studies presented in [Bollen and Hassan, 2011, page 192–197] show that accepting overvoltages during a small percentage of time (less than 1%) will increase the hosting capacity by 50 to 100%. This might actually be the most cost-effective solution, but the risk of equipment damage due to overvoltages will be carried by the other network users, whereas the advantages are with the owners of the distributed generation and with the network operator.

A second option (corresponding to "hard curtailment") is to equip the wind-power installations with overvoltage protection. The overvoltage protection disconnects the production unit when the voltage magnitude is too high. The production is restarted when the voltage is sufficiently below the limit for sufficiently long time. The choice of the reconnection criterion is a trade-off between limiting the amount of time that the production is interrupted and the risk of repeated rapid voltage changes due to connection and disconnection of wind power. This method puts the risk of system overloading (overvoltages in this case) with the owner of the distributed generation; the risk of equipment damage due to overvoltages remains small. When curtailment takes place often, this will impact the amount of produced energy a lot. As shown in Section 4.5 and in Bollen and Etherden [2011] for wind power, the amount of produced energy will reduce when the installed capacity gets too high. However, for moderate amounts of installed capacity, this kind of hard curtailment gives an increase in produced energy.

The overvoltage protection that is part of most anti-islanding protection schemes also takes care of preventing overvoltages by tripping the unit when an overvoltage occurs. The setting for long-term overvoltages (longer than a few seconds) is 110% according to IEEE Std.1547 and 115%

according to EN 50438. Within Europe, the values according to national regulations vary between 106% and 120%. These settings are aimed at preventing equipment damage due to overvoltages due to uncontrolled islanding. The protection also prevents such damages during normal operation. But a too strict setting could limit the amount of distributed generation that can be connected to the distribution system. In the remainder of this section, we will assume that there are other methods available to prevent uncontrolled islanding.

An alternative, instead of tripping the unit completely when an overvoltage occurs, is to limit the production gradually with increasing voltage [Bollen and Hassan, 2011, page 204–209]. The control principle is illustrated in Figure 4.10, where the blue solid curves indicates the relation between the power produced by the unit and the voltage magnitude at the location where the unit is connected to the grid. Changes in production capacity result in the curves moving up and down. Below a certain voltage magnitude, U_{ref} in the figure, the production is not curtailed and the unit produces as much as is possible. For voltages above an upper threshold, U_{max}, the production is curtailed to zero. In between the two thresholds, the production is curtailed. The result of this method is that the voltage will not exceed the upper threshold U_{max} due to the distributed generation.

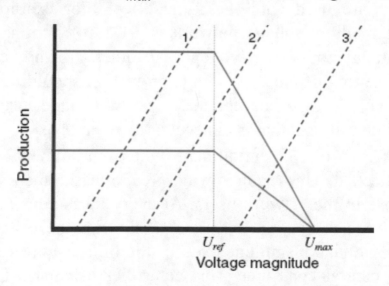

Figure 4.10: Principle for overvoltage limitation: 1 = high consumption; 2 = medium consumption; 3 = low consumption; upper curve: high production; lower curve: low production.

The red dashed curves give the relation between injected power and voltage magnitude. The slope depends on the resistance at the point of connection. The intersection with the horizontal axis is the voltage magnitude without local production. These curves move to the left and the right with varying consumption. In this example, there is no curtailment for high consumption and also not for low production and medium consumption. There is some curtailment for high production and medium consumption, but there is serious curtailment for low consumption. But even for low consumption, the curtailment is much more for high production than for low production.

When setting the overvoltage protection or the curtailment scheme according to Figure 4.10, it should be considered that the voltage magnitude experienced by other network users is not the same as at the terminals of the generator. The situation can occur that the latter voltage is lower than the voltage elsewhere. There is thus an overvoltage for one or more network users, but there is no overvoltage with the generator. Such a situation could especially occur when the generator is connected to the medium-voltage network. The transformation ratio is not the same for all MV/LV transformers. The difference can be up to 5% with the aim to compensate for the voltage drops along a medium-voltage feeder. In some countries, series boosters are used with the same effect. For example, when the voltage in the medium-voltage network is 107% of nominal, it could be 112% of nominal for a low-voltage customer. To prevent this from occurring, the setting of the overvoltage protection should be 5% less than the maximum voltage that is acceptable for a low-voltage customer. But this could result in curtailment that is unnecessarily excessive.

That brings us to soft curtailment, where curtailment is never more than actually needed. This requires widespread measurement of the voltage magnitude close to the network users. At every measurement location, the voltage magnitude is continuously compared with the overvoltage limit. Only when the voltage magnitude exceeds the limit is a message generated and sent to a central controller. This controller calculates the amount of curtailment needed and communicates this to the production units to be curtailed. A control method like shown in Figure 4.7 could be used for this. This curtailment scheme requires a communication infrastructure covering a lot of network users. The actual communication is limited however. The access time requirements of the communication network depends strongly on the overvoltage limits used.

The actual voltage at which curtailment is activated will depend on the overvoltage limits that the network operator has to comply with as well as the immunity of end-user equipment to short-duration overvoltages. There will most likely be a time-voltage relation included, where higher voltages require faster intervention.

4.4.5 RESERVES IN DISTRIBUTION NETWORKS

An important advantage of curtailment has to do with the way in which many distribution networks are designed. Although distribution networks are operated radially, they often have a meshed structure. This allows for spare capacity to be used in case a component is out for repair or maintenance. Repair of overhead lines is relatively quick; the damage is often clearly visible and the line can be accessed easily. However, repair of underground cables and of transformers can take a long time. To prevent that customers are without electricity during these long repairs (days, up to weeks in the worst case), a spare supply path is needed. Serious damage to large transformers sometimes even takes months to repair. Also here, the presence of reserves limits the amount of renewable electricity production that can be connected to the grid.

Consider, for example, the situation shown in Figure 4.11: a medium-voltage distribution network is supplied from two HV/MV transformers. The bus breaker on the secondary side of the transformers is normally open in this example, but the same reasoning holds when the breaker is normally closed.

For each of the two networks, the hosting capacity for distributed generation is roughly the sum of the transformer rating and the minimum consumption. This assumes that transformer overloading is what sets the limit. The hosting capacity is thus $S + L_{min1}$ for the feeders on the left and $S + L_{min2}$ for the feeders on the right. The total amount of distributed generation that can be connected to the medium voltage network is equal to $2 \times S + L_{min1} + L_{min2}$.

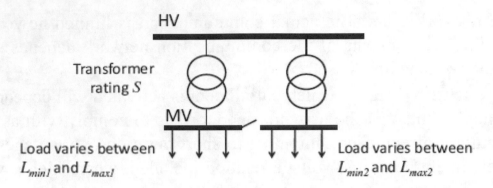

Figure 4.11: Distributed network supplied by redundant transformers.

To allow the whole network to be supplied from one transformer, in case the other is out for maintenance or repair, the maximum amount of distributed generation is only $S + L_{min1} + L_{min2}$. The price of having a high availability is that less distributed generation can be connected. (There is a similar limit on the amount of consumption; we will come to that later.) Using curtailment, the risk of overload can be shifted to the owners of the distributed generation. The reserve is no longer in the form of an additional transformer but in the form of curtailment. The risk for the owner of the distributed generation is that production is curtailed so often that the profitability of the installation becomes too low.

Consider a simple numerical example. The maximum consumption for each partial network is 45% of transformer rating (the total, 90% of transformer rating, can be supplied through one transformer and there is some margin for future load growth); the minimum consumption is 25% of the maximum consumption. Considering reserves, the hosting capacity for distributed generation is 136% of the maximum consumption; without reserves (i.e., using curtailment) about 247% of maximum consumption can be connected before serious curtailment is needed. Thus, 80% more distributed generation can be connected without any new primary infrastructure.

The same reasoning holds for sudden growth of the consumption. The maximum amount of new consumption that can be added, while maintaining the reserve capacity, is $S - L_{max1} - L_{max2}$. With curtailment, i.e., shifting the reserves to the consumption side, this can be increased to $2 \times S - L_{max1} - L_{max2}$. Using the same numerical example as before, the margin for new consumption is increased from 11% to 122% of the existing maximum consumption: thus 11 times as much new consumption can be connected.

One should be aware here that this additional hosting capacity for new production or consumption does not come for free. When the transformer is out of operation for repair or maintenance, the production or consumption might have to be curtailed quite often. It might be able to plan maintenance in such a way that the need for curtailment is limited, but failures, and thus repair, cannot be planned. Repair of a transformer can take several weeks. While a producer can be compensated for the loss on income, curtailment of, for example, electric heating during several weeks might be unacceptable. A risk assessment is needed by the owners of production and new consumption before a decision is made. Curtailment will certainly not always be the preferred solution, but having curtailment available as a design option allows for a wider trade-off between risks and costs.

4.4.6 VOLTAGE AND CURRENT QUALITY

Power quality (voltage and current quality) is normally not considered in the discussion on curtailment. However, also here is to possible to set up a curtailment scheme where curtailment is activated when voltage or current distortion exceeds pre-defined limits. Such schemes appear at least theoretically possible for power-quality variations like harmonic distortion, unbalance and flicker. For power-quality events like voltage dips, rapid voltage changes and transients curtailment schemes would not be of much use; the disturbance would be over before the curtailment would be activated.

Consider a curtailment scheme for the fifth harmonic as an example. The fifth harmonic is the dominating harmonic in most distribution networks and the one that is of most concern for many network operators. Limits could be set for the fifth harmonic current through a transformer or for the fifth harmonic voltage with the network users. When one of these limits is exceeded, curtailment is activated. The selection of which network user or equipment should be curtailed is not straightforward. Different equipment injects harmonic currents with different phase angles. Some currents add whereas others cancel each other. The disconnection of a device injecting a fifth harmonic current may decrease but also increase the harmonic voltage for other network users and the current through the transformer.

In some cases it is known which customer or type of equipment is responsible for the main distortion. Implementation of a curtailment scheme could be for example as follows in those cases:

- When the fifth harmonic current through the transformer exceeds a pre-defined limit, a trip signal is communicated to the customer or the equipment responsible for the main distortion. The customer or equipment is reconnected when the current drops below a lower threshold for a certain time.

- The fifth harmonic voltage is measured with the customers that cause the main distortion. When this voltage exceeds the limit, curtailment is activated.

When it is not known beforehand which equipment contributes to the distortion and which equipment mitigates it, an additional check has to be made before equipment is disconnected. Equipment should only be disconnected when this results in a reduction of the harmonic voltage and in the total fifth-harmonic current through the transformer. Such a check requires an accurate measurement of both fifth harmonic voltage and current, including their angular difference. Together with knowledge of the source impedance for the fifth harmonic current, it is possible to make a decision.

A possible application of a scheme like this would be for an area where large numbers of electric cars, solar panels and heat pumps are installed. To prevent overloading of the grid, the cars are only charged when there is sufficient solar power available and the heat pumps are coordinated as well to further prevent overloading. High levels of harmonic voltage distortion could however occur when solar panels, heat pumps and chargers are all active at the same time. If the probability of this happening is small, a curtailment scheme might be the most cost-effective solution.

4.4.7 ANGULAR STABILITY

Angular instability is a fast phenomenon (a matter of seconds) and it is difficult to predict. Angular instability occurs because the difference in voltage angle between two areas in the transmission systems increases too much during a fault on a major transmission corridor between these two areas. After the fault, oscillations in angular difference occur between the two areas that are normally damped and bring the angle back to a stable value. But if the angle exceeds a certain "critical angle", its value keeps on increasing and the synchronism between the two areas is lost. This, in turn, results in the separation of the two areas and/or the loss of multiple production units in one or both areas. A blackout is often the result. Angular instability is normally prevented by ensuring that faults on major transmission corridors are cleared within a "critical fault-clearing time". During the planning of the transmission system, it is further taken care of that this is even possible when a major line is out of operation. See Bollen and Hassan [2011, page 425–437] for more details on angular stability and how it is impacted by distributed generation.

Angular stability puts a strict limit on the amount of power that can be transported through a transmission corridor: the more power is transported, the lower the critical fault-clearing time. Possible criteria for initiating a curtailment algorithm would be the following:

- Curtailment starts when the power transferred through the corridor exceeds a limit found from simulation studies. This is, in fact, the method currently used by transmission-system operators, with the difference that this is done manually and that only production units are curtailed. An automatic curtailment scheme would not be time-critical because as long as no short-circuit occurs the system remains stable.

- Curtailment starts when the fault-clearing time comes too close to the critical fault-clearing time (the maximum fault clearing time for which the system is stable). The critical fault-clearing time is determined by simulation studies. There are, however, a lot of assumptions and uncertainties in such calculations which will have to be taken care of by using a sufficient margin. This scheme would require very fast reaction because it would only come in place once angular stability is close. This scheme would also require a measurement of the fault-clearing time independent of the protection system. Some of the work on automatic analysis of power-quality disturbances might be used to automatically determine the fault-clearing time from voltage or current recordings [Bollen and Gu, 2006, Section 7.4].

• Based on the value of the angular difference between the two areas and its increase with time, a decision is made on whether the system is stable or not. This requires an even faster reaction than the previous scheme. It also requires the measurement of angular differences (e.g., by means of phasor measurement units) and the fast communication of these to a central point where the decision about curtailment is made. Initial schemes would likely use the angular difference and the rate of change of angular difference as criteria.

Based on existing technology, only the first criterion is available at the moment. More advanced schemes would require the development of simulation methods with sufficient accuracy to limit the risk of angular instability occurring when the second or third criterion is used. Synchronized phasor measurements with a sampling rate of 100 Hz are needed according to Bakken et al. [2011] to make a decision. This puts high requirements on the communication system, especially as the measurements have to be made at locations that might be several hundreds of kilometers apart. As part of the development of simulation methods, the actual angles and fault-clearing times during short-circuit faults should be recorded and compared with the results from simulations.

4.4.8 FREQUENCY STABILITY

Frequency stability is related to the balance between production and consumption in an interconnected system. When a large shortage of production occurs, the frequency will drop quickly and a blackout will result. Frequency instability occurs at a similar time scale as angular instability, several seconds, but frequency instability is easier to predict than angular instability. See Bollen and Hassan [2011, page 422–423] for a brief discussion on frequency instability.

For frequency stability, the overload indicator to be used is the frequency in the system. This is exactly what the existing underfrequency-load-shedding schemes are doing. In some cases, also the time derivative of the frequency ("rate-of-change of frequency" or ROCOF) is used to decide if and how much load has to be disconnected. When using ROCOF as an indicator, it should be kept in mind that during major unbalance between production and consumption, the frequency somewhat differs through the system. The impact on the frequency itself is limited, but the rate-of-change of frequency could be significantly different at different locations in the system.

The easiest implementation of curtailment to prevent frequency instability is to trip the devices taking part in the scheme with underfrequency protection. Instead of tripping complete feeders as part of the underfrequency load shedding, individual devices would be tripped. This would require no communication; the frequency at the terminals of the device could be used as an indicator of the balance between production and consumption in the whole interconnected system.

A more advanced scheme would use a droop line, as used for the power-frequency control. Instead of completely switching off the device, its consumption would be reduced when the frequency drops; the bigger the drop in frequency, the bigger the reduction in consumption. However, for many devices, switching off is easy, but reducing a certain amount is difficult. A frequency dependent temperature setting of electric heating, air conditioning and refrigeration would be a possible solution. But even this will have to be developed and standardized further before it can be used on a larger scale.

When communication between a central controller (or aggregator) and the individual devices is available, the central controller can take care of the setting of the individual devices. Each individual device will still switch off completely at a certain frequency, but the load as a whole will reduce with frequency according to a droop line. The central controller calculates how many devices should be switched off at any moment in time or what would be appropriate droop lines of, for example, temperature versus frequency. These settings are next communicated to the individual devices.

In the existing system, frequency instability is prevented by maintaining sufficient reserve capacity ("spinning reserve") even for the loss of one or two of the largest production unit. If even this should not be enough, the underfrequency load shedding disconnects a substantial part of the consumption to save the system (see Section 4.2). More frequent use of under-frequency load shedding but with a better selection of which consumption or production to disconnect could significantly reduce the need for reserve capacity. Economic studies are needed to decide which devices or installations are most appropriate to equip with such protection. Most challenging probably is to develop a method for the system operator to know how much load-shedding potential (i.e., operating reserve) is available at any moment in time.

4.4.9 VOLTAGE STABILITY

Voltage instability occurs when there is a local shortage of reactive power. Two types of voltage stability should be considered: short-term voltage stability mainly due to large reactive-power demand after a short-circuit fault and long-term voltage stability related to transfer of power over long distances.

Short-term voltage stability is mainly a local issue. A sustained undervoltage after a fault could be the criterion to use. Curtailment could consist of the immediate disconnection of heavy consumers of reactive power. Short-term voltage collapse occurs fast and thus requires a fast intervention. But it might be possible to use only local criteria (voltage and reactive-power flow) to trigger the curtailment. If a sufficiently reliable curtailment algorithm can be developed, this would enable a reduction of the spare capacity in the grid. It could, for example, make it easier to connect wind turbines using induction machines to weak parts of the grid.

An example of curtailment to prevent short-term voltage instability in an industrial installation with large amounts of induction motors is presented by Massee and Rijanto [1995]. A rapid growth in the size of the installation without corresponding enforcement of the supply created a situation where faults at some locations resulted in short-term voltage instability. A curtailment scheme is proposed where different parts of the installation get different settings for their undervoltage protection. The least important parts are tripped first, the most important parts are tripped last.

Long-term voltage stability is a completely different issue. Maintaining sufficient reserve in all cases is the existing way of preventing instability. Long-term voltage instability results always in major blackouts so that it is important to prevent it as much as possible. Also for long-term voltage stability, the limit is the amount of power that can be transported over a major transmission corridor, the same as with angular stability. The amount of active power is, with voltage stability, also strongly dependent on the amount of reactive power transported. Under the existing rules of operation, curtailment is used to ensure sufficient operating reserve: even after the loss of any single component the system should remain stable. This is done manually by the transmission system operator based on simulations to estimate the transfer capacity. This manual scheme could be replaced by an automatic scheme that, for example, limits the amount of power produced by a wind-power installation so as to ensure that there is always sufficient reserve to prevent voltage collapse.

An alternative scheme is to only activate curtailment once the system is on its way to voltage collapse. Long-term voltage collapse is a relatively slow process that could take several minutes. But to detect the onset of voltage collapse, a system-wide monitoring system is needed. In fact, the development of several such systems has started already. Those systems will initially be employed to shed load and/or production as a last resort in case of a major failure in the system. Somewhat further in the future, when such systems become more reliable and more trusted, the amount of reserve could be reduced and instead curtailment could be activated more often. Like with under-frequency load shedding, sufficient curtailable production and loads should be available and the system operator should have information on how much is available.

4.5 CURTAILMENT EXAMPLES

4.5.1 EXAMPLE 1: OVERLOAD DUE TO SOLAR POWER

The first example concerns a hotel with solar power on customer-side of the meter. The amount of solar power that can be connected is limited by the subscribed power of the hotel. The maximum consumption is about 280 kW per phase; the subscribed power is 300 kW per phase. Measurements of the consumption have been combined with a simulation of the solar-power production to estimate the curtailed and produced energy as a function of the installed capacity of solar power. Curtailment is activated once the apparent power exceeds the subscribed power.

The impact on the amount of produced energy from solar power has been calculated for hard curtailment and for soft curtailment. For hard curtailment, the whole solar-power installation is disconnected once the supply current exceeds a threshold. For soft curtailment, the solar power production is not reduced to zero but reduced just enough so that the current does not exceed the threshold. The results are shown in Figure 4.12.

For installed capacity below about 1150 kW, no curtailment is needed, but for higher installed capacity, the need for curtailment increases fast. For an installed capacity equal to 1500 kW, curtailment is needed for more than 600 hours per year. For hard curtailment (red solid curves), the curtailed energy is so much that the annual production decreases with increasing installed capacity. An immediate conclusion is that hard curtailment is not a solution in this case. With soft curtailment (green dashed curves), the amount of curtailed energy is limited, and the annual production continues to increase, but with a decreasing profitability. Up to about 1250 kW installed capacity, the annual production corresponds to 1500 hours/year at peak. The additional 250 kW up to 1500 kW have an effective utilization of only 1300 hrs/year. For the next 250 kW, this would drop further to 1000 hrs/year.

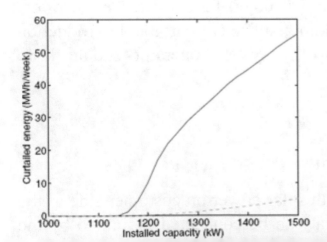
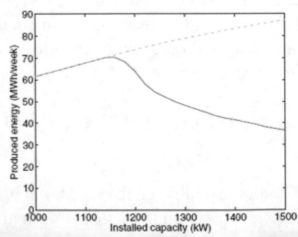

Figure 4.12: Amount of curtailed energy (left) and produced energy (right) for hard curtailment (red solid) and for soft curtailment (green, dashed).

4.5.2 EXAMPLE 2: TRANSFORMER OVERLOAD DUE TO WIND POWER

The second example concerns the building of large numbers of wind turbines on the secondary side of a 19-MVA, 130/10-kV transformer. The maximum consumption downstream of the transformer is about 18MW. Like in the first example, measurements of consumption have been combined with simulations for the wind-power production. The results are shown in Figure 4.13.

Like in the previous example, the impact of two curtailment methods has been studied: hard curtailment where the wind power is disconnected every time, the transformer loading gets too close to its maximum-permissible loading and soft curtailment where the minimum amount of curtailment is used so that the loading of the transformer does not exceeds its limit.

Up to about 26MW installed wind power, no curtailment is needed. For installed capacity up to about 31MW, the total produced energy still increases even with hard curtailment, but the profitability of the capacity above 26MW is rather small. For soft curtailment (green curve), the annual production continues to increase even for high installed capacity. But also here the profitability decreases with increasing installed capacity. Up to 25MW, the utilization is about 2700 hours per year, but the 10MW up to 35MW are only used for 2500 hours/year and the 5MW up to 40MW only for 1900 hours/year.

4.5.3 EXAMPLE 3: OVERVOLTAGE DUE TO WIND POWER

This example studies the use of curtailment to prevent overvoltages in the low-voltage network due to wind power in the medium-voltage network. Measured voltage-magnitude variations were combined with simulations of wind-power production. The measurements and simulations were scaled such that 1MW injected power would raise the maximum voltage to the overvoltage limit at which curtailment is activated. The results are shown in Figure 4.14.

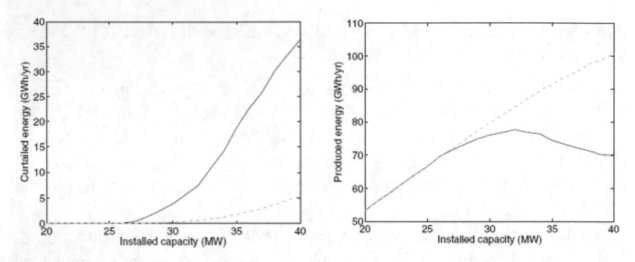

Figure 4.13: Annual amount of curtailed energy (left) and annual energy production (right) for hard curtailment (red solid) and for soft curtailment (green, dashed).

Below 1MW installed capacity, no curtailment is needed; even for 2MW installed capacity curtailment is only needed about 70 hours per year. For higher installed capacity, the need for curtailment increases fast, reaching 2000 hours per year for 4MW installed capacity. For hard curtailment, the amount of curtailed energy grows quickly and for more than 3MW installed capacity the annual produced energy starts to decrease. Under that scheme, it makes no sense to install more than 3MW of wind power.

To illustrate how the local voltage-magnitude variations impact the curtailment, the calculations for overvoltages due to wind power have been repeated for three locations. Three different measurements of voltage-magnitude variations were used, obtained at different locations in different countries and at rather different periods in time. The results of the calculations are shown in Figure 4.15.

What matters for an investment decision is, somewhat simplified, the income in relation to the investment. The income is roughly proportional to the amount of energy produced per year, whereas the investment consists of a fixed part and a part that is roughly proportional to the installed capacity. The ratio between the energy production per year and the installed capacity is a dimensionless parameter; it is referred to as the "capacity factor" of the production unit. When a production unit would produce its rated capacity continuously, its capacity would be 100% or 8760 hours per year.

When comparing between installing, for example, 2 or 3MW, what matters is how much additional energy is produced because of the additional capacity, the "marginal capacity factor". The calculations for these three locations have been used to obtain the marginal capacity factor for both hard and soft curtailment. The results are shown in Table 4.3. The effective utilization is, with values up to almost 4000 hours per year, rather high compared to most locations. The reason is that an average wind speed of 9 m/s has been used in the calculations.

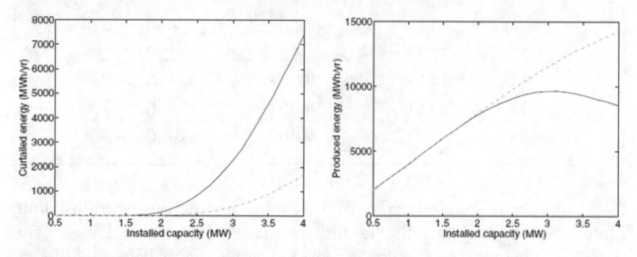

Figure 4.14: Annual amount of curtailed energy (left) and amount of produced energy (right) for hard curtailment (red solid) and for soft curtailment (green, dashed).

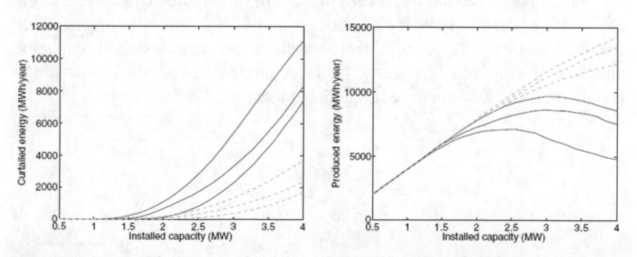

Figure 4.15: Annual amount of curtailed energy (left) and produced energy (right) for hard curtailment (red solid) and for soft curtailment (green, dashed), at three different locations.

Table 4.3: Marginal capacity factor at three different locations in hours/year

Capacity	Location 1		Location 2		Location 3	
(MW)	soft	hard	soft	hard	soft	hard
1.0	3980	3980	3980	3980	3980	3980
1.5	3980	3970	3980	3870	3910	3520
2.0	3950	3710	3820	2770	3660	2230
2.5	3790	2610	3470	1830	3130	520
3.0	3450	1070	3070	850	2520	-
3.5	2940	-	2640	-	1840	-
4.0	2390	-	2140	-	1400	-

From the table, we clearly see the diminishing returns when installing more wind power than the hosting capacity without curtailment. The decrease in marginal production is shown to vary strongly between different locations. As we already observed before, hard curtailment actually reduces the annual production above a certain amount of installed capacity: 2.5MW for location 3; and 3MW for locations 1 and 2. For soft curtailment, there remains some gain in production with additional installed capacity, although the marginal capacity factor reduces quickly after 3 to 4MW installed capacity, again depending on the location.

CHAPTER 5

Market Incentives

In this chapter, we will discuss the different methods in which network users can become active in the various electricity markets. The emphasis will be here on incentive mechanisms for customers to adapt their production or consumption to the needs of the grid. The general term "demand response" is used for that, as different from curtailment as was discussed in Section 4.4. With demand response, any decision to reduce or increase production or consumption is up to the network user to make; other stakeholders (network operators, retailers, etc.) can influence that decision by changing the price or tariff, but the ultimate decision is made by the network user.

We will give an brief description of the main electricity markets in Section 5.1 and Section 5.2. The different forms of demand response are discussed in Section 5.3. In Section 5.4, we discuss the balancing market and customer-participation in that one. Possible structure of markets including congestion at distribution level are discussed in Section 5.5 and ancillary-service markets in Section 5.6. Various developments on the customer-side of the meter are the subject of Section 5.7. In that section, we will discuss how different types of network users can participate in the different markets and curtailment schemes.

5.1 WHOLESALE AND RETAIL MARKETS

Electrical energy as a product is traded on two different markets: a "wholesale market" where the large buyers and sellers trade and a "retail market" where the consumers buy their electricity. There are also balancing markets and ancillary-service markets, which will be discussed on Section 5.4 and Section 5.6. Next to a price for the electricity they consume, network users pay a network tariff for the use of the electricity network.

More about this in Section 5.5. Some producers receive an additional income per kWh next to the price they receive on the electricity market; this could, for example, be incentives for renewable electricity production. This can among others result in negative prices on the wholesale market, see Figure 5.3.

The day-ahead wholesale market, where the price of the electricity is set, is operated as a "spot market": sellers and buyers come together and decide the price at which the electricity will be delivered. In reality, the sellers and buyers do not physically come together at the same spot, but they submit selling and buying bids to a market broker who in turn determines a price based on a set of rules that are known to everybody. The individual bids are however typically not known to everybody. Once the price is set, it is communicated to all players in the market, as well as to which extent their bids have been accepted by the market. As consumption of electricity strongly varies through the day, the market price also varies through the day. A price is typically set for pre-defined intervals the day before the physical delivery of the electricity, intervals range from 15 minutes to 1 hour. The rules of the day-ahead market and the consequences of this will be discussed in Section 5.2.

There remains a certain part of trade in the form of bilateral agreements between a seller and a buyer; the price of these is a matter of agreements between the two players, and there are no rules for this. Also, some consumers, e.g., industrial customers, produce electricity for their own consumption. We will not go into further detail of this.

Only large buyers and sellers of electricity take part in the wholesale market. The large buyers are mainly the so-called "electricity retailers" (or, "electricity suppliers") who buy electricity on the wholesale market and sell it to end-consumers on the retail market. Only a small number of very large consumers buy their electricity directly on the wholesale market. All other customers have a longer term contract with a retailer for delivery of electricity. In many countries, the customer can freely choose and change retailer. The retail market for electricity thus works in the same way as any other retail market. A large part of what is called the "deregulation" of the electricity industry has to do with ensuring that the retail market is as open as possible.

Electricity retailers typically offer their customers a fixed price over a longer period, ranging from one month up to a few years. With monthly prices, the costs of electricity for the consumer follow the seasonal and long-term trends in electricity price on the wholesale market. Here there is again the same trade-off between costs and risks as with many other goods: with variable price, the consumer runs the risk of unexpected price peaks, but long-term costs will be lower because the retailer does not have to add a "risk surcharge". Monthly prices have started to become introduced in a number of countries, but hourly prices remain very rare. As the majority of consumers are currently not exposed to the hourly variations in electricity price on the wholesale market, they do not have any incentive to adjust their consumption to that price. The adjustment of consumption to the wholesale price is called "demand response" and will be discussed in Section 5.3.

5.2 THE DAY-AHEAD MARKET

The main electricity market is the "day-ahead market", also known as the "spot market". This is where the producers, retailers and the largest consumers trade electricity and where the electricity price is set for hourly, half-hourly or quarterly intervals (depending on the country). We will first discuss the principle of price settlement on the spot market and next the ways in which congestion is prevented by means of market principles.

5.2.1 THE SPOT MARKET

The principle of the market settlement on the spot market is illustrated in Figure 5.1. The market broker accepts selling bids and buying bids, for each hour of the day. (Selling bids are also known as offers.) The selling bids indicate the minimum price for which the seller is willing to deliver a certain volume of electricity. The buying bids indicate the maximum price for which a buyer is willing to buy a certain volume of electricity. The buying bids are sorted by decreasing price (the red solid curve in the figure) and the selling bids by increasing price (the green dashed curve). The intersection of the two curves gives the market settlement: the volume of electricity that is delivered and the price at which this is delivered. For more details on the operation of the spot market, see for example Bhattacharya et al. [2001] and Wangensteen [2007].

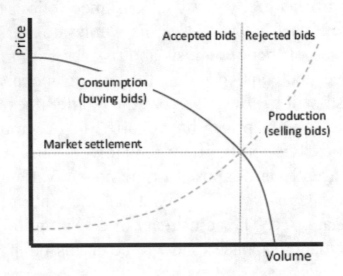

Figure 5.1: Principle of the market settlement on the spot market for electricity.

The market settlement price is such that the volume of buyers willing to sell at this price or lower is equal to the volume of buyers willing to pay this price or higher. Thus, nobody has to pay a higher price than they are willing to pay (according to their buying bid) and nobody has to sell at a lower price than they are willing to sell for (according to their selling bid). Those not willing to pay the settlement price and those not willing to sell at this price will have their bid rejected.

In several U.S. markets, the term "marginal price" is used. The marginal price is the price for producing an additional megawatt of electricity. This corresponds to the market-settlement price in Figure 5.1. There are some differences like the way in which losses and congestion are included (more about this later) and also in the model for the price elasticity of consumption. But in practice, the day-ahead market functions in similar ways in different countries.

There are some properties of the electricity market that have shown to result in price peaks that are hard to avoid. The details vary again between different markets, but the overall properties are often the same. On the selling (production) side there is a certain volume that the owners of the production units are willing to sell at a low price. This includes electricity long time to start, like nuclear power. Next to that, there is a smaller number of units that can be started quickly but with high marginal cost. The owners of these will place high selling bids to the markets. The result is that the selling curve has a long flat part followed by a steep increase.

On the consumption side, the situation is different. As mentioned before, most consumers buy electricity from their retailers at a price that is fixed for a longer period of time. There is thus no price elasticity, and the consumption will not be impacted by the price. This shows up as a vertical line in the diagram of price versus volume. That vertical line moves to the left and to the right with the variations in demand. This is illustrated in Figure 5.2. The result is that the price variations on the wholesale market are mainly due to the differences in selling bids. The buying bids do not have much impact on the market price. In fact, in several wholesale markets, the price elasticity of consumption is not considered at all: in these markets, the price of buying bids is not considered, only the volume of the buying bids (i.e., the "consumption").

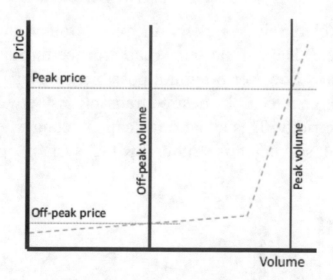

Figure 5.2: Market settlement for peak and off-peak hours in a typical wholesale market for electricity.

The introduction of large amounts of renewable electricity production will sometimes result in negative prices. The owners of renewable electricity production are often willing to bid for a negative selling price, i.e., they are willing to pay for producing electricity. Next to the price for electricity delivered to the wholesale market, the owner of such a production unit has other sources of income related to the amount of kWh delivered. This could be a market for green certificates, payments by a network operator to reduce local losses, or infeed tariffs. As long as the sum of the electricity price and these additional incomes is positive, it is worth producing electricity. After all, the marginal production costs are zero. This is illustrated in Figure 5.3. The production from renewables exceeds the demand, and as a result, the settlement price becomes negative.

The figure also shows the situation without renewables (one may think of a quiet cloudy day). The selling bids from the renewable electricity production are no longer part of the market, and the production curve moves to the left. The result is a higher price on the day-ahead market. This is especially visible for high consumption where the presence of renewable electricity production avoids the price peak. The overall result will be larger variations in price on the day-ahead market.

5.2.2 LOCAL PRICES AND MARKET SPLITTING

Several of the electricity markets in place today are of very large geographical extent. Having the same market price throughout such a large geographical area is however often not possible. The transmission network is not sufficient to enable all the desired transactions between the sellers and the buyers of electricity. This situation is called "congestion", and there are different ways of solving this, either by the system operator or by the market.

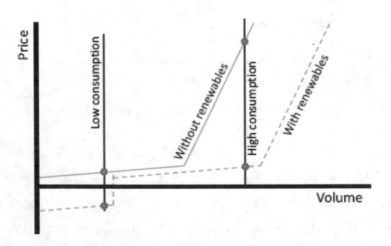

Figure 5.3: Market settlement during low and high consumption (red vertical lines), with (green dashed curve) and without (green solid curve) a high amount of production from renewable sources.

With so-called "market splitting", different prices hold for different price areas. As a first step, the market always finds a settlement assuming there is no congestion, (the "copper plate assumption"); the resulting price is called the "system price". If the settlement for the whole market would result in congestion, the market is split into different price areas. The price will increase in areas with a shortage of production and decrease in an area with a surplus of production.

Several North American markets use the "location based marginal price" (LBMP), this is the cost to serve the next megawatt of consumption at a certain location. The costs due to losses are included in the LBMP. Without losses and without congestion, the LBMP would be the same anywhere in the system; including losses gives some spread in the prices; congestion can result in prices that are much higher at one location than at another. The LBMP is the sum of the system price, the marginal cost of losses and the marginal cost of congestion. The marginal cost of losses can be negative, zero, or positive. The main difference between market splitting and LBMP is that LBMP will automatically result in a different price at different locations, whereas market splitting requires a decision from the system operator to split the market. Another difference is that market splitting does not include the costs of losses whereas LBMP does.

A method for keeping the price the same throughout the market is "counter trade", where the system operator pays generators in the "surplus area" to reduce production and pays generators in the "shortage area" to produce. The costs of this counter trade are charged to all the network users through the network tariffs. The market is in this case not impacted by the congestion, but the costs of congestion are spread over all network users. Counter trade gives an incentive to the transmission system operator to reduce congestion, but removes any incentive for location-based demand response.

An example of the variation in marginal price is shown in Figure 5.4 and 5.5, for 15 zones in the market operated by the New York Independent System Operator on the 1 August 2011. The marginal price itself during the day is shown in Figure 5.4. The price rises during afternoon hours when the temperature, and thus the consumption, is highest. But most strikingly is that the differences in price between different locations become very big. Whereas the highest price is "only" 65% higher than the lowest price at one location, it is almost seven times as high at another location.

The impact of losses on the price is relatively small as shown in Figure 5.5; the marginal price of losses might lead to an increase but also to a decrease in price. The impact of losses is biggest in the afternoon hours when the power flows are biggest and the system is heavily congested. The degree of congestion is clearly visible in the right-hand graph of Figure 5.5 showing the marginal price of congestion. Only in some exceptional cases is the marginal price of congestion negative, in almost all cases does this result in an increase of the electricity price. As shown in the figure, serious congestion starts for one area at 11 am, but for most areas, the system is seriously congested only between 2 pm and 6 pm.

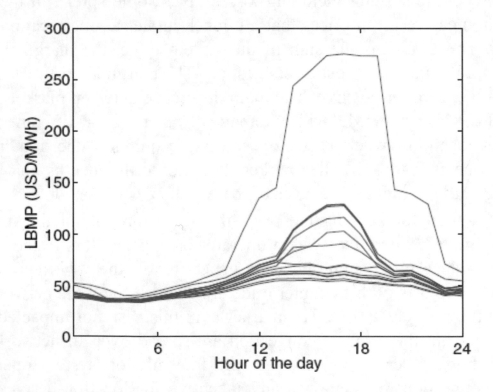

Figure 5.4: Variation in marginal price during one day for 15 different zones.

In several European markets, the contribution from wind power has already become so big that negative spot prices occur. An example, for

Denmark, is shown in Figure 5.6. The system price (upper curve) remains rather constant around 60 Euro per MWh. There, is however, a surplus of wind power in Denmark that cannot be transported to the rest of the system due to congestion. A result is that the price in Denmark becomes much lower and during some hours even negative.

5.3 DEMAND RESPONSE

When we talk about "demand response", we refer to a customer adapting its consumption to the electricity price. Normally, the term is reserved for reducing consumption during peak prices only or for shifting consumption from hours with high price to hours with low price. It is in this meaning that we will use the term here. An overall reduction in electricity consumption because of high electricity prices is normally not considered in the discussions, but this is demand response as well.

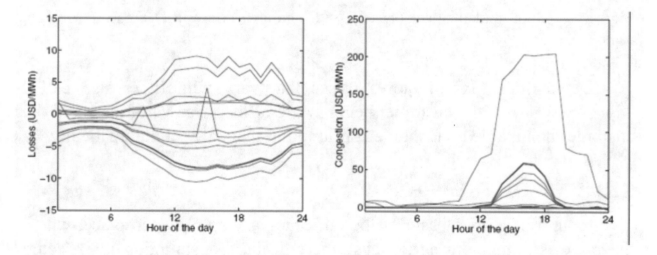

Figure 5.5: Variation in marginal price of losses (left) and congestion (right) during one day. Note the difference in vertical scale.

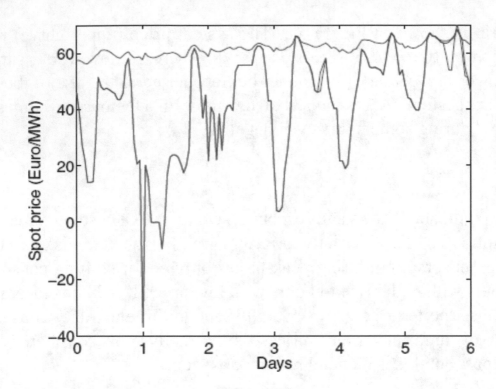

Figure 5.6: Spot-market prices (system price and area price) in Denmark during six days.

Demand response requires prices that vary with time, also called "dynamic prices", the customer being aware of the price variations, and the customer being able to adapt the consumption manually or automatically.

5.3.1 TIME-OF-USE PRICING

Most countries have a basic method of time-varying prices in place, where the price is higher during periods where a high consumption is expected ("peak price") than during the rest of the year ("off-peak price"). The peak-price and off-peak-price periods are defined based on the time of the day, time of the week and time of the year. For a country with peak consumption during summer, the peak-price period would typically be the afternoon hours on working days during summer. When the peak occurs during winter due to electric heating, the peak-price periods would typically be 8 am to 8 pm on working days during winter. Some network operators and retailers use three price periods. The details are very much dependent on the local variations in consumption. An example, by the Ontario Energy Board, is shown in Table 5.1.

Table 5.1: Example of time-of-use price

	May - October	November - April	Weekends
Off-peak	7:00–19:00	7:00–19:00	00:00–24:00
Mid-peak	7:00–11:00 and 17:00–19:00	11:00–17:00	
On-peak	11:00–17:00	7:00–11:00 and 17:00–19:00	

The change in price is typically communicated to the customers by means of power-line communication signals. These same signals can also be used to switch equipment on and off; for example, water heaters could be switched on at 7 pm and off at 7 am. A time-off-use price often results in a consumption peak immediately after the start of the low-price period.

Whereas this method of pricing has been in place for many years in many countries and has served its purpose well, there are reasons for moving a step further. The most important technical reason is that although consumption is on average higher during the peak-price periods, it still varies a lot even within the peak-price period. On some days, an additional reduction in consumption would be desirable. The worst case is when one or more important production units or transmission lines are out of operation during a period of extremely hot or cold weather. When this margin between consumption and production capacity is smallest, the risk of a blackout is biggest.

A second reason for moving beyond the simple time-of-use price is that the incentive to reduce consumption during peak prices is, in fact, rather small. The ratio between peak-price and off-peak price varies between utilities, but a factor of two is about as high as it gets. This has shown to provide insufficient incentive to reduce consumption. A sharp increase of the peak price would however impact the total electricity bill too much and not be acceptable to the consumers.

A third reason is the introduction of new types of production and consumption. New loading patterns will appear, several of which will no longer be linked to the time of day, week or year as before. Network and system loading will become less predictable, with a result that time-of-use pricing become less useful to reduce peak loads.

5.3.2 OTHER METHODS

Different more-advanced methods of dynamic pricing are being investigated and in some cases already introduced. These methods roughly fall into the following categories, but the variations within the categories are rather big, see for example Braithwait [2010].

- The consumer pays a price that is linked to the hourly price on the day-ahead wholesale market. A small margin between expected consumption and production capacity will result in a high price on the wholesale market. This high price will give the consumer an incentive to reduce consumption; the smaller the margin, the higher the price, and the higher the incentive. The hourly prices are settled the day before so that they can be easily communicated to the consumer. The consumer can next reduce consumption during hours with high price.

- The normal fixed prices hold during most of the days, but when a small margin between consumption and production capacity is expected, the price is increased significantly, for example a factor five. The consumer is informed about this before, typically on the afternoon of the day before after the closure of the day-ahead market. The consumer can again reduce consumption during the hours with a high price. This method is called "critical peak pricing". In most experiments and existing schemes, the critical peak price is fixed for a whole year, but even more dynamic pricing schemes are possible.

- An alternative scheme is where the consumer does not pay a higher price but instead receives a payment for reducing consumption during certain hours. Again the consumer is informed about this ahead of time, typically in the afternoon of the day before. This is called "critical peak rebate". Like with critical-peak prices, a variable rebate could be introduced based on the actual need for demand response during a given hour.

- The consumer pays a price for electricity that is directly linked to a local real-time price on the wholesale market.

• The consumer is, through an aggregator, part of the balancing market.

5.3.3 HOURLY PRICING

When a substantial fraction of the consumers gets exposed to the hourly price, this will give an incentive to them to reduce consumption during periods with high prices. In market terms, there will be price elasticity that will especially impact the price peaks. How this matters to the day-ahead market is shown in Figure 5.7: below a certain price, the consumption is not impacted. For higher prices, the consumption will be reduced, which shows up as a diagonal part in the curve built from the buying bids. However, certain customers are not exposed to hourly prices and have therefore no incentive to reduce electricity, so below a certain volume, the curve will go vertical again.

During low-price periods, there is no incentive to reduce consumption so that the off-peak price will not be impacted by the demand response. But the peak price will be reduced: the bigger the price elasticity, the more the price will be reduced. When more and more consumers are exposed to the hourly price, the price elasticity will be bigger and the peak price will get lower. Here it should be noted that the consumers are only exposed to the wholesale market price through their retailer; the consumers are still not a part of the wholesale market. In Figure 5.7, it has been assumed that the retailer is fully aware of the price elasticity of the consumers. This will obviously not be the case, especially not in the early stages of the introduction of demand response. The consequences of this will be discussed in Section 5.4.

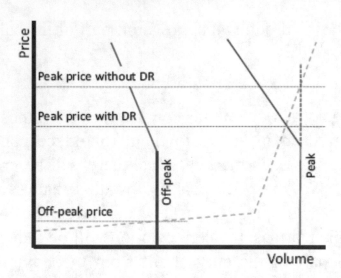

Figure 5.7: Impact of demand response on the electricity price during peak and off-peak periods.

One of the types of demand response studied during the PowerCentDC demonstration project [PowerCentsDC, 2010] was to offer consumers hourly prices. Those prices were based on the actual hourly prices on the day-ahead wholesale market operated by PJM. Prices were available on the project website, by calling a toll-free number and were communicated to the smart thermostats that some of the project participants were equipped with. The participants were directly notified, via telephone, SMS or email, ahead of days with high prices.

5.3.4 CRITICAL PEAK PRICING

At this moment, critical peak pricing (CPP) is the method that is most in use already (France, Florida and San Diego are examples). This is also the method that is most studied in demonstration projects.

In France, consumers with a subscribed power of 6 kW or more can opt for a time-of-use price. Electricity is about 50% more expensive during peak hours. About 30% of consumers in France, corresponding to 60% of consumption, are exposed to time-of-use prices [Badano, 2010]. On top of that, consumers with a subscribed power above 9 kW and that remain part of the regulated market, can opt for the "Tempo scheme". During each year there are under this scheme 300 days with normal prices, 43 days with medium prices, and 22 days with high prices. The days with medium and high prices occur between November and March and are directed mainly towards consumers with electric heating. The price is announced via internet, e-mail and text message (SMS) at 17:30 the day before. An indicator lamp with the customer lights red on days with high prices, white on days with medium-level prices, and blue on days with normal prices. The days are referred to as red, white and blue days, respectively. This scheme has been in place since 1996 and about 300 000 domestic customers and 100 000 small commercial customers take part. The total reduction of the national consumption is about 150MW on days with medium prices and about 300MW on days with high prices [Badano, 2010]. The prices during the six periods, relative to the price averaged over the year, are as follows:

- On normal days: off-peak-price 54%; peak price 68%.

- On days with medium prices: off-peak price 117%; peak price 140%.

- On days with high prices: off-peak price 223%; peak price 637%.

The Energy Select program by Gulf Power in Pensacola, Florida, was introduced in 1998 and has about 10 000 participants. The program contains four price levels: low (68% of the annual average price); medium (82%); high (135%) and critical (377%). The low, medium, and high prices form a normal time-off-use scheme: the dates and times are fixed with high prices between 1 and 6 pm in summer and between 6 and 10 am in winter. The critical price is in place up to 87 hours per year, one to three hours at a time, and is announced one hour before. The customer is informed by means of an indicator lamp in the home. Participation in the program comes together with a programmable thermostat that automatically reacts to price signals obtained from the utility.

San Diego Gas and Electric offers a critical peak prices scheme to larger customers (subscribed power of 20 kW or more) that are equipped with a remotely-read meter with 15-minute interval reading. On top of a normal time-of-use tariff with three price levels during winter and three different ones during summer, a maximum of 18 so-called "CPP events" can be called. The customer will be notified by 3 pm the day before; the price during such an event is about 10 times the normal peak price.

In the PowerCentsDC demonstration project [PowerCentsDC, 2010], critical peak price was one of the three different incentives for demand response that were tested. During the demonstration project, the critical peak prices came in place based on the temperature predictions: above 90°F (32°C) in summer; below 18°F (-8°C) in winter. Customers were informed of the occurrence of critical peak prices the day before, by 5 pm in the demonstration project. Critical peak prices were five to six times higher as the normal prices and where always in place during four hours (afternoon hours in summer; early morning and early evening in winter). During the demonstration project, there was no direct link between the activation of critical peak prices and the actual loading of the system. The activation was based on the temperature prediction only. There were 12 days per year with critical prices in summer and 3 such days in winter.

Another experiment was conducted in California during 2004 and 2005 in which consumers were offered different time-dependent electricity prices above the existing multi-tier pricing scheme. The following three schemes were offered:

- "Time-of-use pricing", where the price was higher during a pre-defined peak period. The prices varied with time of day, time of week and time of year, but they were determined beforehand. The prices were not impacted by the actual loading of the network.

- "Critical peak pricing - fixed" where much higher prices were in place during up to 15 days a year. The timing was unknown beforehand but consumers were informed the day ahead.

- "Critical peak pricing - variable", where the notification time could be as short as 4 hours and the critical time period up to 5 hours. The customers were offered the technology to automate their demand response.

5.3.5 CRITICAL PEAK REBATE

With "critical peak rebate", the electricity price remains the same, but instead the consumer receives a payment in return for reducing consumption during certain critical peak hours. The critical peak hours can be determined and communicated to the consumer in the same way as for critical peak prices. To know how much a consumer reduced consumption compared to not taking part in demand response, a baseline consumption has to be decided. Determining this baseline is actually the weak point of critical peak rebate, but it is a necessary part of any such scheme. A detailed discussion on methods to estimate the actual demand reduction is given by Goldberg [2010]. Some examples of methods to determine the baseline, as summarized in PowerCentsDC [2010]:

- PJM: average consumption during the same hours of the three days with the highest consumption out of the ten previous weekdays without critical peaks.

- New York Independent System Operator: the five highest out of the ten previous weekdays.

- Anaheim public utilities: the three highest days during the first half of summer.

- San Diego Gas and Electric: average of the previous five non-event, non-holiday weekdays.

In the PowerCentsDC demonstrator, the three highest days in the billing month were used as a reference. For normal domestic consumers, the peak rebate used within the project was between 2.5 and 5 times the price for consumed electricity, higher in summer than in winter.

An additional correction to get a fair baseline for consumers would be to multiply by a certain factor to account for the on average higher consumption of an individual customer on days with a higher overall consumption. From the data of the 2005 Anaheim demonstration project, it was concluded that consumption of individual consumers not part of a demand-response scheme on critical peak days was on average 23% higher than for the same hours of similar weekdays. This would mean that the baseline should be placed 23% above the similar-day average.

5.3.6 MULTI-TIER PRICES

A special type of price structure that is in place among several utilities is the one where the price per kWh consumption increases above a certain monthly or annual consumption. The sole aim of this is to encourage consumers to reduce their total consumption. The system is commonly in use in the United States. An example is the four-tier rate for San Diego Gas and Electric (effective from 1 April 2011):

- Up to 100% of baseline: 14 cent/kWh.

- 101 to 130% of baseline: 16 cent/kWh.

- 131 to 200% of baseline: 29 cent/kWh in summer; 27 cent/kWh in winter.

- Above 200% of baseline: 31 cent/kWh in summer; 29 cent/kWh in winter.

A future extension would be to increase the price based on the amount of carbon-dioxide emission linked to a customer's monthly consumption. Under a scheme like this, the carbon-dioxide emission is calculated on an hourly basis from either the marginal production or the production during that hour. This would be an incentive to reduce electricity consumption during hours with high carbon-dioxide emission.

5.3.7 REDUCTION IN PEAK CONSUMPTION DUE TO DEMAND RESPONSE

An overview of the effectiveness of demand response programs for domestic customers is given by Braithwait [2010]. The reduction in peak consumption for the different types of demand response is as follows (in percent of the pre-reduction peak):

- 10 to 30% for critical peak pricing.

- 20 to 50% for critical peak pricing in combination with technology to

- 10 to 20% when customers are exposed to hourly wholesale prices.

- up to 33% when hourly wholesale prices are combined with technology to automatically reduce consumption.

Two interesting conclusions are the following: critical peak pricing appears to be more effective than hourly pricing. The effectiveness of the demand response can be increased a lot by using automation to reduce consumption with high prices. The effectiveness of demand response is also dependent on the climate conditions. The fixed CPP price used during the California experiment, resulted in a demand reduction of around 15% in southern California and around 9% in the north, where temperatures are lower.

Only by making use of the existing demand response program, the peak demand for the U.S. could be reduced by 4% by the year 2019. Extending these programs to the whole country would result in a reduction of 9% [Hamilton, 2010].

The peak reductions observed in the PowerCentDC project [PowerCentsDC, 2010] are summarized in Table 5.2. It shows that hourly prices do not result in a significant reduction in consumption. This may however be due to the fact that high prices, comparable to the critical peak prices and rebates, only occurred twice during the whole experiment. Critical peak prices result in a substantial reduction in consumption, more so in summer than in winter, despite the peak prices being rather similar. Critical peak rebate appears to provide less incentive although the rebate is, at least during summer, about the same as the critical price.

Table 5.2: Measured demand response during the Washington DC experiment

Incentive scheme	Reduction in peak	
	Summer	Winter
Critical peak pricing	34%	13%
Critical peak rebate	13%	5%
Hourly prices	4%	2%

The effectiveness (or, "price elasticity") for industrial customers varies between industry types. The most energy-intensive industries are most responsive to high prices. Based on studies done with two different utilities, about one third of industrial and commercial customers shows substantial demand reduction during price peaks and about a same amount shows no reduction at all [Braithwait, 2010].

5.3.8 DEMAND RESPONSE AT SYSTEM LEVEL

Demand response programs are offered by a number of large system operators, including PJM, the New York ISO, the ISO New England, and the Midwest Independent Transmission System Operator. These program do not directly involve the electricity consumers but demand response aggregators, also called "curtailment service providers". Such an aggregator can place a bid for a certain amount of demand reduction into the wholesale market in the same way as a generator would place a bid for a certain amount of production. When the bid is accepted, the aggregator is committed to reduce consumption by the agreed amount. To obtain a certain amount of demand response, the aggregator in turn has contracts with its customers to reduce consumption either through price incentives or compulsory. In the latter case, some kind of signal is needed to switch off certain loads on the customer premises.

For example, PJM offers a number of demand-response services, mainly through demand-response aggregators, referred to as "curtailment service providers". A distinction is made between three different types of demand response:

• Compulsory emergency demand response. An aggregator can sign up to the capacity market with a certain capacity and receive a capacity payment in the same way as a generator. Upon request from the system operator, the demand has to be reduced by up to the agreed capacity. The resources must be available for up to 6 hours per day, up to ten days during summer. The demand response aggregator is treated in the same way as a generator, with the difference that the aggregator has to rely on its customers to provide the demand response.

- Voluntary emergency demand response. Even an aggregator that did not sign up to the capacity market can place a bid when the system operator requires additional capacity. The aggregator will receive a payment for the demand response but no payment from the capacity market.

- Economic demand response, which will be activated on a voluntary basis when the price on the day-ahead market is high.

An example of the way in which aggregation works is a program offered by Baltimore Gas and Electric to its customers. In return for a fixed compensation per month, this utility is allowed to switch off air conditioners and water heaters with participating customers. Baltimore Gas and Electric has about 300 000 participating customers and uses this to bid 600MW on the wholesale market operated by PJM [Hamilton, 2010]. We see the interesting situation here that demand response at transmission level is obtained by curtailment in the retail market.

5.3.9 EXAMPLE - SWITCHING OFF CONSUMPTION

As an example, the variation in hourly price on the wholesale market (the "spot price") is shown in Figure 5.8 for Sweden during 2009 and 2010. A logarithmic scale is used for the vertical axis to be able to show the large price spikes at the end of 2009 and the beginning of 2010.

The general pattern shown is that the spot price stays within a rather narrow band most of the year, with some hours of very low prices. In fact, the spot price was zero for a number of hours in the early morning of 26 July 2009, which was not accidentally the period with the lowest consumption of the year. Very low prices also occur in spring when the hydropower reservoirs are filled by the melting snow. During the winter period, when the consumption is highest, the spot price is overall higher, but it also shows a number of very high price peaks. These price peaks offer the biggest opportunity for consumers to save money, assuming that their price is linked to the hourly spot price and that they have the ability to cut consumption or to move consumption to hours with a lower price.

Assume that a consumer is willing to pay up to a certain price for electricity. When the spot price exceeds this limit, the consumer will reduce consumption. The lower the price the customer is willing to pay, the more hours during the year that the consumption will be reduced. This is shown in Figure 5.9, based on the spot prices from Figure 5.8. For a willingness-to-pay of 400 Euro per MWh or more, the demand response is limited to a few hours per year; below 200 Euro per MWh, the amount of hours with demand response increases quickly.

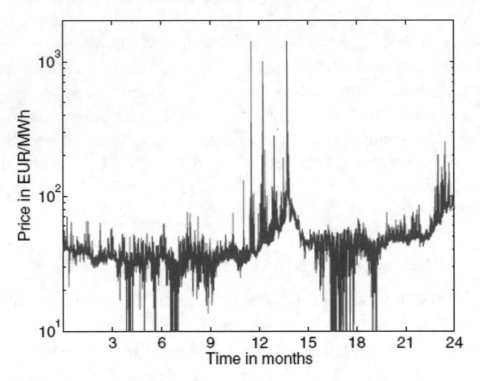

Figure 5.8: Variation in spot price during 2009 and 2010. Note the logarithmic vertical scale.

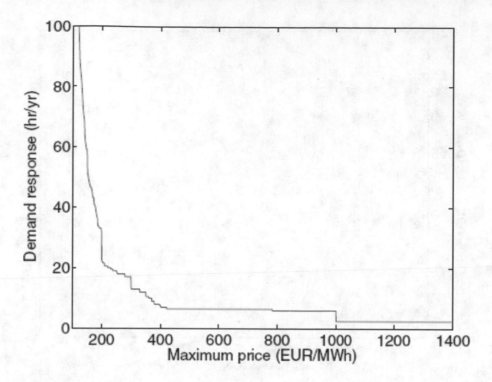

Figure 5.9: Amount of hours with demand response as a function of the maximum price a customer is willing to pay.

Although the consumption is only reduced during a few hours per year, these are the hours with the highest electricity costs, so that the savings could be substantial. The savings on the annual electricity bill are shown in Figure 5.10. Here it is assumed that the part of the consumption that takes part in the demand response is the same throughout the year. The higher the price the customer is willing to pay, the higher the annual electricity bill. For a willingness to pay above 1400 Euro per MWh, there are no savings at all because the spot price never gets higher than this value. When the customer is not consuming electricity when the spot price is above 400 Euro per MWh, the savings are about 7400 Euro per year for each MW of consumption taking part in the demand response.

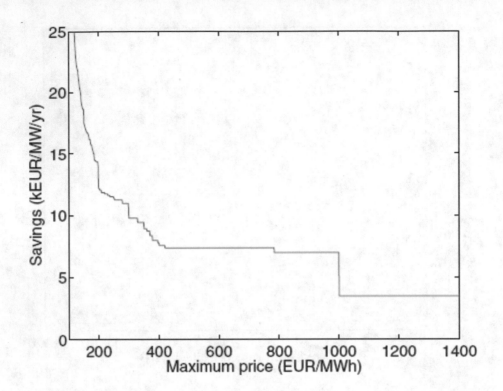

Figure 5.10: Annual savings in electricity costs per MW of participation in demand response, as a function of the maximum price a customer is willing to pay.

The savings in costs obviously come at the expense of having to reduce consumption. This is illustrated in Figure 5.11. The horizontal axis shows the amount of time during which demand response is required; the vertical axis shows the savings in percent of the annual electricity bill. By switching off a certain load during 1% of time, the annual costs of electricity can be reduced by 5.5%.

From the curves in Figure 5.9 and 5.10, we can get an impression of the amount of money that a customer can save through demand response. From Figure 5.9, we can observe that when the customer is willing to pay less than 200 Euro per MWh (20 Eurocent per kWh), the number of hours during which the load should be disconnected increases very quickly. For most customers, this will be too inconvenient. From Figure 5.10, it follows next that the savings amount to 15 000 Euro for a load of 1MW. For an industrial installation with several MW of load that can be turned off 20 hours per year, the savings can be substantial. However, for a domestic customer with only a few kW of load that can be part of the demand response, the savings are some tens of Euro. It is important to note that the control and communication needs associated with demand response are about the same for a domestic customer as for a large industrial customer.

5.3.10 EXAMPLE - SHIFTING CONSUMPTION

As a second example, consider the case where electricity for a certain process or device is only needed during a limited number of hours every day, where it doesn't matter when during the day. In case the customer is exposed to the hourly spot price, the hours with the lowest price can be used. This will save costs compared to using electricity during the more expensive hours of the day.

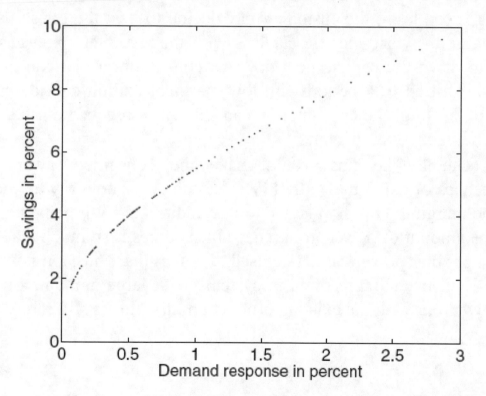

Figure 5.11: Savings as a percentage of annual electricity costs, versus percentage of time with demand response.

To illustrate the gain, the annual electricity bill is calculated for two cases using the spot prices. In the first case, the electricity is used during the hours with the lowest spot price. In the second case, the electricity use start at 8 in the morning and continues for as long as it takes. The ratio of the costs for these two cases is shown in Figure 5.12.

For processes that only require electricity a few hours every day, the costs can be reduced by about 30%. Having hourly prices and the ability to shift consumption thus allows big savings. When electricity is only needed during working days (Monday through Friday) the gain is even bigger, up to 37%.

5.3.11 DEMAND RESPONSE BASED ON CARBON-DIOXIDE EMISSION

A specific type of demand response is based on the marginal global carbon-dioxide emission due to consuming an additional kWh. When the marginal emission is too high, the customer may decide to heat the house using gas or to charge a hybrid-electric car less with the risk that some gasoline is needed to drive the car the next day. All consumption that can be shifted should be shifted to a period with low marginal carbon-dioxide emission. Instead of shifting, the consumption can simply be reduced during the hour without being recovered later.

A scheme like this would make the consumers aware of the consequences of using that extra kWh. It is however not easy to know how much the marginal emission is due to one additional kWh being consumed. When the amount of power from renewable sources is so much that some of the wind or solar power cannot be used, the marginal emission is zero. This situation will occur more often in systems with large amounts of solar or wind power, but with the existing production mix, this is still rarely the case in most countries.

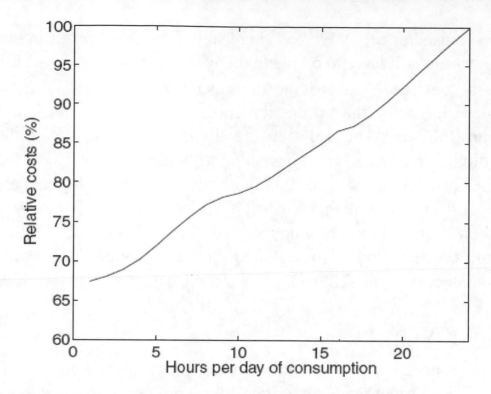

Figure 5.12: Relative costs with demand response compared to no demand response for consumption that is only needed a limited number of hours per day.

When the total available production capacity from wind and solar power can be absorbed by the grid, there are two ways of calculating the marginal emission. One method is to consider the marginal production on the day-ahead wholesale market. It is this source that will produce more when the consumption is higher. This assumes however that the consumption is known beforehand. If the customer only decides at the last moment to use or not use a certain amount of energy, it will be the marginal source on the balancing market that is impacted. The marginal emission would be the emission from this marginal source.

The situation is again different when a substantial part of the production comes from hydropower or when hydropower is used for balancing. We talk about hydropower using dams and reservoirs; flow-of-river is similar to wind and solar power. In a hydropower-dominated system, the marginal emission based on balancing power will often be zero, also there might be situations where the marginal emission based on the day-ahead wholesale market would be zero. But the energy in the reservoir can only be used once; an increased consumption could result in a shortage later that would still have to be made up by electricity from fossil fuel.

An alternative is to provide the customer with the average carbon-dioxide per kWh for the production mix from the day-ahead market for each hour of the day. This will not actually have much impact on the global carbon-dioxide emission, yet it has an important educational value. A large-scale use of a scheme like this will also result in higher electricity consumption and thus higher electricity prices during periods with many low-carbon sources in the production mix. With reference to Figure 5.3, the right-hand curve will shift to the left in the case "without renewable" and the left-hand curve will shift to the right in the case "with renewables".

5.3.12 RECOVERY PEAK

A serious concern with demand response programs based on day-ahead pricing, like hourly pricing or critical peak prices, is the occurrence of a recovery peak once the price drops again to a more normal level. A lot of reduction in demand is, in fact, shifting the demand to a later moment in time. This holds especially for heating and cooling load and has, for many years, been recognized as a concern with rotating interruptions. A plot of the consumption by a small customer with and without demand response is presented by Goldberg [2010]. Without demand response, the peak consumption is about 3.3 kW. Demand response during six hours reduces the consumption during the original peak period to 2 kW, but in return it results in a new peak of about 3.9 kW during the two hours after the demand response was active. The net result of demand response is thus that the peak consumption increased from 3.3 to 3.9 kW. In Kiliccote [2009], it is stated that "Negative demand savings are often found after a demand-response period as part of a recovery peak in which the HVAC or cooling system tries to bring the thermal zones back to normal conditions."

The recovery peak has also been observed during a demand-response experiment conducted in Norway during 2007 [Sæle and Grande, 2011]. A clear reduction in consumption was visible during periods with high prices, but the recovery peak after those periods resulted in a higher hourly consumption than without demand response. The maximum hourly consumption for the customers participating in the scheme increased from 3.5 kWh to 5.2 kWh.

5.4 BEYOND THE SPOT MARKET

5.4.1 OVERVIEW OF DIFFERENT MARKETS

As was mentioned before, the spot market is a day-ahead market, and after market closure, the bids are binding and the spot price is fixed. However, the consumption and production of electricity cannot be fully predicted, while at the same time production and consumption have to be equal during the actual operation of the system (the "physical delivery" as it is called in market terminology). To take care of the deviations between the predictions used for the spot-market settlement and the actual production and consumption, a number of additional markets are in place. These are all occasionally referred to as "balancing markets" although they are of very different character and function. The structure of these markets varies between countries and regions, with different terminology being used in different countries; see for example the overview of balancing markets in different European countries by Verhagen at al. [2006]. In general, the following markets can be distinguished:

- The "intraday market" where trading takes place between the closure of the day-ahead market and the moment of "gate closure". This is where corrections can be made based on new information that has come in. Prediction of wind and solar-power production can be corrected when new weather information becomes available. That weather information might also result in revised predictions for consumption from electric heating (in winter) or air conditioning (in summer). Also, a non-planned outage of a production unit or a major transmission line might result in intraday trading. The intraday market can have many buyers and sellers.

- The "balancing market": up and down regulation after gate closure, typically under the control of the system operator. The market operator gathers bids for "up regulation" and "down regulation". With an up-regulation bid, a producer commits to increase production beyond the spot-market volume upon request from the system operator. With a down-regulation bid, a producer commits to reduce production below the spot market volume upon request from the system operator. The balancing market, can have many sellers but only one buyer, which is again typically the system operator.

- The "power-frequency control": automatic adjustment of the production controlled by the frequency on the grid. It is this power-frequency control that keeps the actual balance between production and consumption. Like with the balancing market, there can be many sellers, but there is only one buyer: again the system operator. Sellers place bids for capacity to automatically increase or reduce their production, so called "primary reserve". Once these bids are accepted, the changes in production are taken care of by the automatic control system. There is a payment for providing primary reserve, as well as a settlement afterwards for the net produced energy.

- The "balancing settlement" for deviations in volume compared to the bids into the spot market. In some cases, market participants get paid for deviations, for example, when consuming less or producing more, but, in general, there is a penalty on deviating from the bid. The aim of this settlement is for the system operator to recover the balancing costs from those market participants that caused the unbalance. The settlement rules differ between countries.

In this book, we reserve the term "balancing market" for the up and down regulation under the second bullet above. Several U.S. market operators do not operate a balancing market but a real-time market where a real-time location based marginal price is set every five minutes. The function of this real-time market is the same as the balancing market under the second bullet above.

5.4.2 EXAMPLES OF PRICE VARIATIONS

The price variations on the Swedish balancing market have been used here as an example to illustrate some of the properties of this market. The balancing market in Sweden results in two prices: an up-regulation price and a down-regulation price. Normally, one of the two is equal to the spot-market price. When the actual consumption is more than the one according to the spot-market settlement, additional production is needed. The transmission-system operator buys this additional production against the up-regulation price from those producers that are willing to increase production at short notice. The price is determined in the same ways as the price at the day-ahead market by ranking bids by price. The price for all balancing power is determined by the bidding price of the last bid that has to be accepted to cover the required balancing power. Those producers willing to increase production typically get paid a higher price than the price on the day-ahead market. The positive (red) curve in Figure 5.13 gives the difference between the up-regulation price and the spot-market price during the whole of 2009 and 2010. In general, the difference is small, but some high peaks are visible in the up-regulation price. This corresponds to hours with a shortage of production units willing to increase their production at short notice.

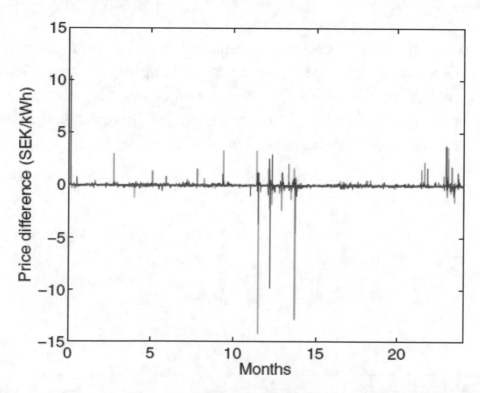

Figure 5.13: Variation of the price on the balancing market during a two-year period.

When the consumption is lower than according to the spot-market settlement, there is a surplus of production. Some production units will have to reduce their production. For this, they get paid the difference between the spot-market price and the down-regulation price. When there is a need for down regulation, the down-regulation price is always lower than the spot-market price. The negative (blue) curve in Figure 5.13 shows the (negative value of) the difference between the spot-market price and the down-regulation price. It is this amount that a producer gets paid when reducing production at short notice.

Two zoom-ins of the curve are given in Figure 5.14. The graph on the left shows a one-month period in 2009 when the balancing price was close to the spot-market price. With some exceptions, the differences were less than 0.1 SEK/kWh. It also follows from this figure that the balancing market "changes direction" regularly, sometimes a few times a day, but the direction can also remain the same during a few days.

In the graph on the right, the price differences are shown for a period, during early 2010, with strong price variations. During some hours, the up-regulation price is high; during other hours, the down-regulation price is very low. Occasionally, the down-regulation price becomes negative. This happened especially during spring when the melting snow resulted in a surplus of water in the big hydropower reservoirs. The occurrence of a negative price has in itself no meaning. What matters is the difference between the spot price and the down-regulation price.

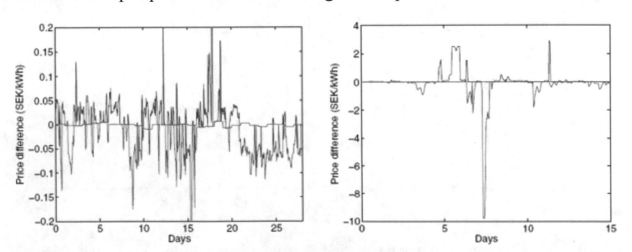

Figure 5.14: Variation of the price on the balancing market during a month with small

difference in vertical scale.

The volume of the balancing market in Sweden is shown in Figure 5.15, where the upper (red) curve gives the volume for up regulation and the lower (blue) curve the negative of the volume for down regulation. During some hours, there is up as well as down regulation, but this is normally not the case. The figure shows that the volume of the balancing market shows a big variation with more than 1000MW occurring several times per year for up-regulation as well as for down-regulation. The main cause of these high values are prediction errors in the consumption.

5.4.3 BALANCING BY DEMAND RESPONSE

In the text above, it was assumed that the balancing power is only provided by producers. But it can equally well be provided by adjusting the consumption. Consider the situation that the consumption is less than the production. Instead of reducing the production, the system can also be balanced by increasing the consumption. The price paid by a consumer for additional consumption would, in that case, be equal to the down-regulation price, which in some cases would be negative. A consumer that only occasionally needs electric power and that has the ability to shift consumption several hours or even days could buy on the balancing market and make use of the lower down-regulation price. During some hours, one would even get paid for using electricity. The disadvantage is, however, that it will only be known during the actual delivery hour whether a bid will be accepted or not. That makes planning difficult, but with some kinds of processes, that is not a concern. Several studies have been done where the charging of electric cars is adjusted to the need for balancing power.

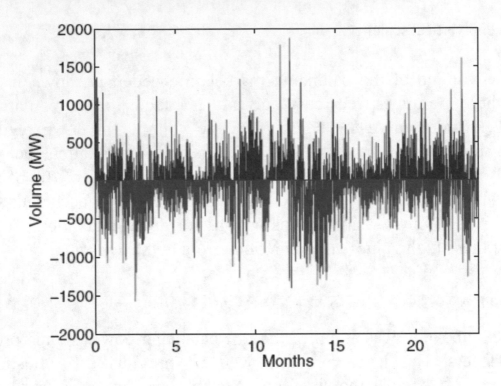

Figure 5.15: Volume of the balancing market during a two-year period.

When there is a shortage of production, this can be compensated by reducing consumption, i.e., by contributing to the up-regulation market by means of demand response. The money obtained for consuming less than a certain baseline will, in that case, be the difference between the up-regulation price and the spot-market price.

Another approach to earn money on the balancing market, which is already being used by some owners of pumped-storage installations, is to buy electricity when the down-regulation price is low and to sell this electricity when the up-regulation price is high.

5.4.4 IMPACT OF DEMAND RESPONSE ON BALANCING

In Section 5.3, we discussed how demand response could result in a reduction of peak prices. The main idea behind demand response is that consumers get exposed to the hourly prices on the spot market and, with that, get an incentive to reduce their demand as a response to price peaks. But as was also mentioned earlier, the vast majority of consumers is not part of the wholesale market but instead part of the retail market. The settlement on the wholesale market is based on selling and buying bids, which are binding to the market participants. There are, however, no buying bids on the retail market; a customer can decide its consumption without any consequences. Still the retailer has to estimate the consumption by its customers to be able to make a bid on the spot market. The retailer thereby has to assume a certain price elasticity. This price elasticity will not be exactly known, especially not when demand response is new, and as a consequence of this, demand response will impact the balancing market.

Consider two cases: price elasticity is overestimated and price elasticity is underestimated. What this matters for the consumption is shown schematically in Figure 5.16 where curve 1 is the price elasticity as used for the spot-market settlement. It is this price elasticity that determines the spot price, independent of what the actual price elasticity is. Any deviation between predicted and actual consumption is taken care of by the balancing markets.

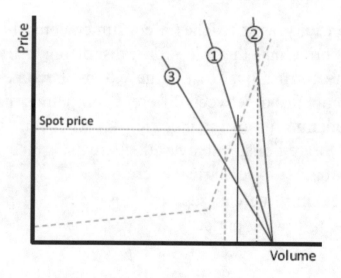

Figure 5.16: Impact of price elasticity on the consumption: 1. price elasticity used for the spot-market settlement; 2. actual price elasticity when the price elasticity is overestimated; 3. actual price elasticity when the price elasticity is underestimated.

When the price elasticity is overestimated (curve 2 in the figure), the demand reduction will be less than expected and thus the consumption higher (the right-hand dotted vertical line in the figure). This additional consumption above the spot-market settlement will have to be provided by the balancing market. It is expected that demand response will start to make a difference initially when price peaks occurs. Price peaks correspond with a shortage in production capacity. The market for upward regulation may be very tight as well, which will result in high prices on the balancing market. The result is that the spot price has been reduced not by actual demand response but by expected demand response. The spot-market price is not impacted by this, but instead, a high balancing price results. The latter is borne by the electricity retailers, not by the consumers that are exposed to the spot-market price.

The other way around, when the price elasticity is underestimated (curve 3 in Figure 5.16), the spot price is too high and the consumption lower than expected. The latter requires a large volume of down regulation on the balancing market. The costs of this depend strongly on the production mix available.

5.5 MARKETS AND THE NETWORK

The price of electricity as it reaches a consumer consists basically of two parts: the energy price and the price for transporting it from production to consumption. The former has, to a large extent, become an open market where there is competition between different companies that produce or sell electricity. The opening up of the electricity market remains incomplete in certain countries or regions, but overall, there is a clear trend towards a completely open market for electrical energy.

Roughly speaking, we can separate the electricity price as paid by a consumer into the following parts:

• Energy costs

• Congestion and transport costs at transmission level

• Use of network tariff for the distribution network

• Government surcharges and taxes

Although we separated those four contributions here, this separation is not always visible to a consumer. Some governments are, for example, creating incentives for renewable electricity production, which could be embedded in the energy costs, in the use of network tariff, as a fixed charge per consumer or per kWh, or paid by taxes. In the same way, congestion costs could be part of the energy bill (e.g., the marginal costs of congestion that increase the location-based marginal price) or be part of the use of network tariff (e.g., counter trade).

The ongoing developments and the vast majority of the research being conducted concerns customer participation in the energy markets, e.g., by means of the different types of demand response. The costs of congestion at transmission level are included in this whenever principles like market splitting or location-based pricing are used. At distribution level, customers still pay a certain amount per kWh, independent of when they use their electricity; next to that, many distribution network operators charge a fixed amount per customer per year, typically based on the subscribed power and the voltage level at which the customer is connected. Distribution network operators also charge a "connection fee" when a new customer is connected or when a customer makes a major change in its installation that may impact the network.

As we saw in Section 5.2, there are marked mechanisms in place to prevent congestion at transmission level. There are, however, no such incentives in place at distribution level. Instead, the distribution network is assumed to have sufficient capacity to cope with the demands of the market. This is, in fact, often the case, for a number of reasons. Building new distribution lines or cables takes much less time than building transmission lines (one or two years for distribution versus five to ten years or longer for transmission) so that congestion can easier be prevented. The transmission system is continuously monitored from a control room, where overload situations can be detected and mitigated once they occur. At distribution level, this is not the case: the transmission system is operated; the distribution system is built and should function by itself after that. Other reasons for this difference include the much higher reliability requirements in the transmission system and the fact that it is the transmission system that has become most exposed to the opening of the electricity market.

Recent developments however make that, also at distribution level, the need for market mechanisms has arisen. Both new production (wind and sun, domestic combined-heat-and-power) and new consumption (heat pumps, electric cars) could require large investments in distribution networks. Local demand response based on price incentives could be a cost-effective alternative.

The possibilities of creating price incentive by means of the distribution tariffs are basically the same as for the price of energy as discussed in Section 5.3, for example, hourly prices, critical peak prices and critical peak rebates. There are, however, some implementation issues with that, mainly related to the small size of a local market and the unpredictability of consumption and production.

Some of the trends with network tariffs are worth mentioning here before we discuss the more advanced options. The network tariff consists of three parts, where some network operators use only one part, others use two and some use all three.

- The "subscription fee" is a fixed monthly charge for being connected to the network. This part of the tariff typically depends on the subscribed power: the higher the subscribed power, the higher the subscription fee.

- The "energy part" which is a fixed amount per kWh. This could be the same throughout the year or depending on the time of use. On one hand, we see more network operators introducing time of use tariffs. This should give incentives on consumers to reduce their peak load; the costs of infrastructure are strongly dependent on the peak load. On the other hand, we see a political pressure to completely remove the energy part of the network tariff. A recent communication from the European Council stated clearly that future network tariffs should no longer be dependent on the volume of energy transported. As long as the income for a network operator is dependent on this volume, it will not be in the network operators' interest to encourage energy conservation by their customers. By removing the energy part of the tariff, this barrier against energy conservation is removed.

- The "peak-load part" of the tariff creates an incentive to reduce peak load without the income of the network operator being dependent on the volume of energy transported. This has started being used by several network operators in Sweden for medium-sized customers, and two network operators have introduced it for domestic customers; more are looking into the possibilities. Such a tariff requires interval reading of the electricity meters.

In a distribution market, the network tariff would depend on the actual loading situation of the network. None of the above-mentioned parts of the network tariff gives any incentive to adapt consumption or production when the local distribution network is actually heavily loaded. The discussion on possible structures for a distribution market has not even really started, but let us give an example of what such a market could look like.

The customer pays an annual fee for using the network. Next to this fixed part, the customer pays a variable part for the marginal costs made when actually using the network. The variable part of the tariff depends on the loading situation in the local network when the electricity is used. The customer pays to, or receives from, the network operator a certain amount per kWh produced or consumed. The network tariffs for production and for consumption can be positive as well as negative, as we will see later. The network tariffs are independent of any payments to or from a retailer for energy produced or consumed, which are also per kWh but are ruled by developments on the wholesale market and possible congestion in the transmission system.

The distribution tariff for each hour will be different for four different loading situations:

- **Consumption exceeds production; no congestion**. The marginal costs to the distribution network operator for consumption or production are only the losses. Losses increase when consumption is increased and decrease when production is increased. Consumption thus pays a certain price per kWh and production receives this amount per kWh.

- **Consumption exceeds production; network congested**. The local distribution network becomes a price area by itself. A local bidding mechanism, preferable real-time, could set the local price and the difference between this price and the wholesale price would be the network tariff. An alternative construction would be for the distribution network operator to buy electricity locally to prevent overloading. This can be from local production units or in the form of demand response. The price for this will be charged to the local consumers. When there is insufficient local production and demand response available, the local price for consumption could increase a great deal.

- **Production exceeds consumption; no congestion**. We are, in fact, back to the first case: the marginal costs to the distribution network operator for production or consumption are only the losses. But as there is now a surplus of production, the losses increase with increasing production. As a consequence, the consumer gets paid and the producer has to pay a network tariff to cover the losses.

- **Production exceeds consumption; network congested**. The situation is now just the other way around as in the second case: the price to be paid by the producers (and paid to the consumers) will increase until there is sufficient reduction in production and increase in consumption so that the local network is no longer congested.

The price for losses would be the actual losses during the hour in the distribution network times the actual local price of electricity on the wholesale market. The use of network tariff is thus equal to the price on the wholesale market times the percentage of losses in the local distribution network during the hour. When the local wholesale price is low, the costs of losses is low and the network tariff is low. When the wholesale price is negative, as could occur in some markets for a large surplus of renewable electricity production, the costs of losses would become negative. A local producer would have to pay in this case when there is a shortage of production locally. This might sound counter-intuitive at first because there is a shortage of production locally so that there should be an incentive to increase production and reduce consumption and not the other way around. However, as there is no congestion in the distribution network, the local shortage of production does not matter to the distribution network. What matters instead is the global surplus of production and shortage of consumption.

A market mechanism for preventing congestion in the distribution network has been successfully tested during an experiment on the Olympic Peninsula in Washington State [Pratt, 2008, Snowdon et al., 2010]. A simplified version of the price-volume curve that has been used in the experiment is shown in Figure 5.17. The green dotted curve represents the costs made by the distribution network operator. The first horizontal part represents the feeder from which power is obtained. Up to the first knee in the curve, there is no congestion and all demand can be supplied over the feeder. A number of gas turbines are available that can be started when consumption increases beyond the capacity of the feeder. Producing electricity by these gas turbines is more expensive than the location marginal price at transmission level. Starting of the turbines thus results in an increase of the electricity price. The second horizontal part of the green dotted curve represents the costs of producing electricity by the gas turbines. Once they have reached their capacity, the price for using the network increases again sharply and there is no longer any limit.

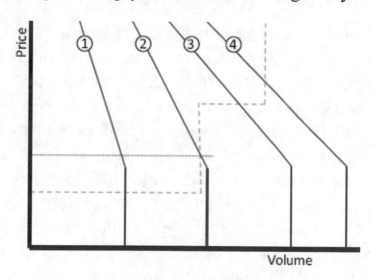

Figure 5.17: Principle of the distribution market as used in the Olympic Peninsula experiment.

The four red solid curves represent the willingness of the consumers to pay. As part of the experiment, consumers were equipped with "smart thermostats" that would compare a set price with the price on the market. They only operated when the temperature was outside of a certain range and when the market price was less than the price the customer was willing to pay. The slope in the curves represents the demand response due to the smart thermostats. For curve 1, the demand is small and all power can be supplied over the feeder. Variations in demand result in the curve shifting in horizontal direction to the left and to the right. The result will be a change in volume of electricity delivered but no change in price. With increasing demand (curve 2), the feeder becomes congested and demand response is activated. This is done automatically; there is no need for human intervention. Variations in demand will now result in variations in price whereas the volume will remain constant. When the demand increases even more (curve 3), the price the consumers are willing to pay becomes sufficiently high to justify the operation of the gas turbines. The price becomes constant again and the volume varies with changing demand. Finally, (curve 4) also the gas turbines have reached their capacity and the price starts to vary again at a constant volume.

5.6 ANCILLARY-SERVICE MARKETS

The large production units contribute to the network in more ways than only injecting electrical energy that can next be transported to the consumers. All these additional ways are referred to as "ancillary services". The shift from large thermal and hydro units to smaller and renewable sources of production will also impact the availability of several of these ancillary services. Where there occurs a shortage of ancillary services in the current system, the system operator ensures that sufficient large units remain connected to the grid. The term "must-run production" is sometimes used for this. Furthermore, instead of this, the ancillary services can be obtained from other providers on an open market. The market mechanisms for some ancillary services are well understood and some markets are already in place. For other ancillary services, the discussion has not started yet and implementation may be a long way away. In this section, we will give a brief overview of the ancillary services and the way in which they could be obtained through a market mechanism. In almost all cases, an ancillary service market will be a market with just one buyer.

5.6.1 OPERATIONAL RESERVE

As was mentioned several times before in this book, the high reliability of the transmission system is based on keeping sufficient operational reserves. Currently, operational reserves are almost exclusively in the form of large production units and additional transmission lines. The result is a certain amount of production and transmission capacity that will be very rarely used or never at all. The costs for building and maintaining that capacity will however be present all the time and will have to be paid one way or the other from the tariff or electricity price.

Shifting the reserves to the consumption side might be more cost effective. See Section 2.5 for a discussion on reserves. Needed for this shift is a bidding mechanism in which consumers place bids for reducing their consumption on demand. Production units could bid on the same market with production bids. Large consumers could take part in an operating reserve market directly, small consumers through an aggregator.

Consider for example a small city supplied via three overhead lines with 250MW capacity each. As long as the maximum consumption from the town is less than 500MW, the three lines offer a secure supply. Exceeding the 500-MW mark will not immediately result in an interruption, but when one of the three lines fails, the other two will be overloaded. Instead of building a new line, the operating reserve can be in the form of demand response. With a peak demand of 550MW, the network operator will have to obtain sufficient offers to reduce consumption by 50MW when ordered by the network operator so as to keep the maximum consumption below 500MW.

5.6.2 FREQUENCY CONTROL

Automatic frequency control (also known as "power-frequency control") is used by the system operator to maintain the actual balance between production and consumption in the power system. In most countries, two different types of automatic frequency control are in place. Primary control acts on a timescale of seconds and ensures the balance between production and consumption over the whole interconnected system. Secondary control is slower, covering timescales up to several minutes, and ensures the balance over each control area. Secondary control is also used to bring the system frequency back to its nominal value (50 or 60 Hz) after the loss of a large production unit. (In control-theory terms: primary control only has a proportional, "P", element whereas the secondary control also has an integrative, "I", element).

Especially primary control requires the availability of sufficient reserves (typically referred to as "primary reserves" contrary to the "secondary reserves" required for the secondary frequency control). The frequency control market is therefore often structured as an operating-reserve market. Once a bid on this market is accepted, the bidder is obliged to automatically contribute to the frequency control. Settlement can be for maintaining the reserve, for activating the reserve, or for both.

Many countries already have a functioning frequency control market (it was therefore already mentioned in Section 5.4), but in most countries, this service is still only provided by conventional production units. There is, however, also here no technical reason for not allowing consumers and small producers to be part of the frequency control. For small production, the implementation would be the same as for large production units: the production would be a function of the frequency in the system. For the nominal frequency, the production would be equal to the production set point. There are, however, some issues to be resolved before small and renewable electricity production can be used on mass for frequency control. For renewable units, there is often no such thing as a "production set-point" that can be used as a reference. Instead, the set point will have to be calculated, for example, from the wind speed. Another, potentially more serious, issue is that contribution to frequency control requires operating reserves. If renewable production units will be used for frequency control, their production will be less than it could have been. Unless there is a surplus in production from renewable sources, this will have to be covered by other sources, possibly fossil fuel.

When consumption is used for frequency control, the choice of the reference point (the consumption for nominal frequency) is again the main design issue. This might not be very important for the stability of the system but could seriously impact the settlement. A customer could get an unexpected payment or bill based on random variations in consumption that were not considered in the calculation of the reference consumption. When frequency control is delegated by individual devices, the choice of the reference point could be easier. For a thermostat, the temperature setting could be made dependent on the frequency; for a battery charger, the charging current could be adjusted.

5.6.3 REACTIVE POWER AND VOLTAGE CONTROL

Reactive power occurs when voltage and current waveforms are not exactly in phase. Explaining reactive power in physical but non-mathematical terms is very difficult if not impossible and no such attempt will be made here. The reader is referred to a textbook on power systems or circuit theory instead.

Next to the flow of energy ("active power") that is always from production to consumption, there is a flow of reactive power that could be in either direction. Reactive power is not associated with any energy consumption itself, but transporting reactive power results in real energy losses and contributes to voltage-magnitude variations. Reactive power is consumed by several types of electrical equipment, especially by induction motors. Lightly loaded overhead lines and underground cables produce reactive power, cables much more than lines. Reactive power is consumed by transmission lines when they are normally or heavily loaded; also loaded transformers consume reactive power. Like active power, also reactive power has to be in balance. The net production or consumption of reactive power is supplied or absorbed mainly by large production units. Also capacitor banks produce reactive power; synchronous condensers and certain types of power-electronic devices can produce and consume reactive power on demand.

The network operator typically requires large production units to contribute to the reactive power balance. The reactive power flows at transmission level are strongly related to the voltage control. The reactive-power balance is mainly held by means of automatic voltage control of the busses to which the production units are connected. Next to that, the transmission system operator does reactive-power dispatch and ensures that there is sufficient reserve also where it concerns reactive power.

There are, at the moment, no markets for reactive power in place, but several authors have proposed such markets. Bids on such a market could be made by the owners of large production units (the traditional p   viders) but also by consumers or even by new players on the market. An issue that would have to be solved before introducing a reactive-power market is the ownership of reactive-power compensation equipment, like capacitor banks, by network operators. An important property of any reactive-power market is that it would be a rather local market; reactive power cannot be transported over long distances without heavy losses and increasing the risk of instability. The result is that there will likely only be a small number of competing market players where a single player can easily dominate the market. This could be a serious barrier against a workable reactive-power market.

At distribution level, the consequence of reactive power flow is mainly an increase in losses, but it also impacts the voltage control. Many network operators do charge their customers for excessive consumption of reactive power. In a future market, network users could be charged or paid based on the marginal costs of losses due to their reactive power consumption or production.

A voltage-control market could be set up at distribution level based on how much a network user contributes to the voltage rise or voltage drop. As long as the voltage is within a certain range, there is no market activity. However, when the voltage is higher than the permitted range, customers that reduce the voltage will get paid whereas those that cause the voltage to rise will have to pay. When the voltage is below the permitted range, it is the other way around. There are similarities and even overlap between such a voltage-control market and the network market discussed in Section 5.5. A functioning network market may make the voltage-control market superfluous. When there is an overvoltage or undervoltage, this can be treated as congestion. The advantage of a separate voltage-control market is that it could create an incentive for network users to contribute to the voltage control. That would however require that the network market and the voltage-control market do not interfere with each other.

5.6.4 SHORT-CIRCUIT CAPACITY

One of the consequences of replacing conventional generation by new sources could be a reduction of the short-circuit capacity at transmission

When the short-circuit capacity gets too low, the transmission network will have to obtain short-circuit capacity from somewhere. The existing solution would be to intervene in the market and keep a number of large production units in the system. A future solution could be a market for short-circuit capacity.

At this moment, there are no minimum requirements on short-circuit capacity, but several national regulators plan to introduce such requirements. Where there are no such requirements, it might be more appropriate for a network operator to set up separate markets, for example, for voltage quality and stability.

5.6.5 VOLTAGE QUALITY

Market arrangements to maintain sufficient voltage quality have been discussed regularly within the power-quality field and several proposals can be found in the literature. A trading system for emission permits is proposed by Driesen at al. [2002], Yang et al. [2006]. In other proposals, the basic idea is that the customer that causes a certain disturbance, for example, harmonic distortion, has to pay, and that the one reducing that distortion will get paid. In the same way as with congestion in the distribution market, the network tariff would be increased for polluting customers when the level of voltage disturbances would become too high.

Consider as an example the fifth harmonic, the one that is often the main concern for distribution-network operators. When the fifth-harmonic voltage would exceed for example 5%, the network tariff would be adjusted for each customer with a non-zero fifth harmonic current. The amount of adjustment would be based on the magnitude and the phase angle of the fifth harmonic current. If this current would increase the magnitude of the fifth harmonic voltage, the customer would have to pay a higher network tariff. Customers with a fifth harmonic current such that it reduces the magnitude of the fifth harmonic voltage would pay a lower network tariff.

There are, however, a number of unresolved issues with the implementation of such a market. One of them is that it requires accurate measurements of the harmonic current and detailed information about the system before the calculations can be made. Even with a trade in harmonic emission, there are measurement issues because of the verification needed

An additional issue not addressed in the studies on such markets is that the total emission is always less than the sum of the individual emissions. This cancelation effect is strongly time dependent and very difficult to predict.

5.6.6. BLACK-START

The transmission system has an extremely high reliability, but occasionally things go wrong and a blackout is the result. Once a blackout occurs, restoring the supply is difficult and time consuming. It can take several days before the supply is back for all customers. An important reason for this is the limited availability of black-start generation. Most large production units require the presence of the grid to start production; at the same time, the transmission system requires certain ancillary services (mainly reactive power) from the production units. Things get even more complicated because many large production units take a long time to start once they have stopped. When the grid disappears, many production units also automatically stop as there is nowhere for them to send their electricity to. Nuclear power stations can take several days to get started again. Some large power stations, including nuclear power stations, have the ability to quickly reduce production upon loss of the grid and to go into "island operation" where they only supply the consumption needed to keep the unit in operation.

To be able to restore the transmission system after a black-out, a transmission-system operator needs access to a number of units with so-called "black-start capability". Also there should be a sufficient number of units with the ability of island operation to limit the time it takes to restore the system. There are normally no power stations that are kept only for their black-start or island-operation capability.

There are, however, costs associated for the production units with having these capabilities, whereas they are only needed once every ten years or less. Also here a market arrangement could be a cost-effective solution. This is especially the case because several new forms of production can provide black start and island-operation capability much easier than large units. It has even been discussed to, after a blackout rebuild the system starting at distribution level instead of starting at transmission level as is currently common practice.

A market for black-start capability would be like an operating reserve market but with even less utilization of the reserve. The payment should thus be predominantly for having the reserve available. An important issue is for the network operator to verify that offered black-start capability will indeed be available when needed.

5.6.7 ISLAND OPERATION OF THE GRID

Controlled island operation of parts of the distribution system was discussed as a type of microgrid in Section 3.5.3. Such "controlled island operation" is also important when rebuilding the power system bottom-up. The network operator does normally not have access to production units and will thus have to rely on others to provide the services to operate part of the grid as an island. A market mechanism could be set up to provide this service; such a market would again be set up like an operating-reserve market. It seems reasonable to assume that controlled island operation will only be set up by the network operator when interruptions are common, so that the reserve would be activated rather often and/or for long periods of time.

Payment of the island-operation service can be based on the availability, on the activation or on both. Like with all reserve markets, a mechanism should be in place to verify if the capability to operate in island is indeed sufficiently available. It may also be decided to charge a fine to market participants that are not able to provide the service when ordered to do so.

5.6.8 INERTIA

Another one of the consequences of the introduction of new sources of production is that the total amount of inertia present in the system may drop substantially. Many modern wind-power installations and distributed generators do not contribute to the system inertia. The consequences of this are increased risks of frequency instability and angular instability [Bollen and Hassan, 2011, page 417–422].

Providing sufficient inertia without access to a large rotating mass is technically challenging but not impossible, and, in fact, several proposals are discussed in the literature for providing artificial inertia using the power-electronics interface with wind turbines and microgenerators. Once that technology is available, setting up a market for inertia is rather easy. In fact, there is no real difference between a market for frequency control and a market for inertia. In both cases, the control system only needs a local parameter (the frequency) as an input, and in both cases, the location of the market participants does not matter. The difference is mainly in the control algorithms needed: these need to be much faster with inertia than with frequency control, but the required speed is not a concern for modern power-electronics converters.

5.6.9 STABILITY

In the power system, stability is maintained in a number of ways. The most commonly-used method is simply to keep the operational point of the system sufficiently far away from any instability; this is known as operating reserve and has been discussed several times before in this book. There is, however, a number of methods in use to keep the system stable even when there is insufficient reserve and many more are proposed in the literature. An example where there are already measures in place is the use of "power-system stabilizers" to damp "inter-area oscillations". Inter-area oscillations occur due to interaction between the controllers of large power stations in different parts of a large interconnected system. The frequency of these oscillations is of the order of several seconds and their damping is often very small. Therefore, a number of large power stations is typically equipped with an additional control system, the "power-system stabilizer", that provides additional damping for these oscillations.

The power-system stabilization function can also be provided by power-electronic converters that are present for other purposes anyway, for example, the converters in modern windparks. Further research and development is also ongoing on using these power-electronics converters to provide damping during subsynchronous resonances and to prevent or delay angular and voltage instability.

Once the technology is available to mitigate instability, the network operator may decide to buy stability services, for example, from wind reduction in operating reserves and thus an increased transport capacity through the transmission system.

5.6.10 OVERVIEW

An overview of the different ancillary-service markets discussed above is given in Table 5.3. The comparison of the markets is based on the following criteria:

- Can the service be provided from anywhere in the system or only at locations close to where the service is needed?

- Is the service often activated or only in rare occasions?

- Can the service be activated and controlled using local measurements only or is a communication infrastructure needed?

- Is this service needed at many locations throughout the power system or only at a limited number of locations?

Table 5.3: Future ancillary-service markets

Service	Anywhere?	Often activated?	Local measurement?	Needed everywhere?
Operational reserve (generation)	Yes	No	Yes	Yes
Operational reserve (transmission)	No	No	No	Yes
Frequency control	Yes	Yes	Yes	Yes
Reactive power (transmission)	No	Yes	No	Yes
Voltage control (distribution)	No	Yes	?	No
Short-circuit capacity	No	No	?	No
Voltage quality	No	Yes	No	No
Black-start	Yes	No	Yes	Yes
Island operation	No	No	No	No
Inertia	Yes	Yes	Yes	Yes
Stability	No	No	Yes	No

These criteria, next to the economics and the availability of the technology, determine if a market will develop in the future or not. Assuming the technology is available the markets that are most likely to develop are the ones with most often "Yes" as an answer to the above criteria. One of the two markets with four times "yes" is the one for frequency control, which is actually already in place in many countries but in most cases limited to large production units.

5.7 NETWORK USER

The earlier parts of this and the previous chapter were mainly presented from the viewpoint of the grid. In this section, we will move the viewpoint to the network user. A general network user and its different interfaces with the outside world (the grid and the different electricity markets) is described in Section 5.7.1. Next, the options for participation are discussed for different types of network user: small consumers in Section 5.7.2, small and large production in Section 5.7.3 and medium and large customers in

Section 5.7.5. Storage is the subject of Section 5.7.4. Some special types of network users are discussed in Section 5.7.6 (electric vehicles) and in Section 5.7.7 (microgrids and virtual power plants). Finally, the communication between the network user and the grid is discussed in Section 5.7.8.

5.7.1 GENERAL NETWORK USER

The various interactions between a general network user and the grid are shown systematically in Figure 5.18.

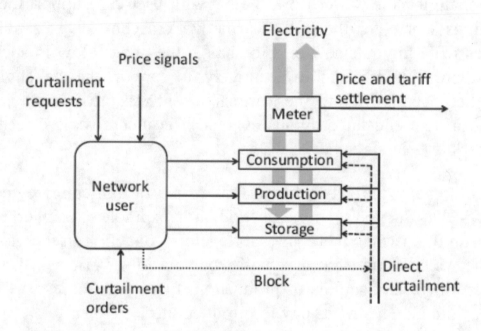

Figure 5.18: The network user and the grid: power flows and communication.

In the general case, the network user has access to consumption, production and storage, all of which can be controlled. The control can be manual (in which case, the "network user" in the figure is a person) or automatic (in which case, a "network-user control system" takes over). The figure shows the general case: most network users have access only to consumption or production, although there is a slow trend for consumers to also have some production on their side of the meter. This is expected to grow a lot in the future and as mentioned in Chapter 2, it is one of the driving forces behind the smart grid. The ownership of storage is rather rare among network users, with the exception of pumped-storage installations.

But here also a trend is expected in the growing use of storage on the customer-side of the meter.

The physical interaction of the network user with the grid is in the form of the flow of electrical energy (the "power flow"): either from the network user to the grid (production) or from the grid to the network user (consumption). At any moment in time, it is only one of the two: either production or consumption. Even over every time interval (five minutes, one hour, one year, etc.), the network user is either a net producer or a net consumer.

By changing the power flow, the network user can support the grid or make things worse for the grid. The network user receives a number of incentives from the outside world to change the power flow in such a way that it supports the grid, more accurately, to support the stakeholder that creates the incentive. How the various incentives are created has been discussed in this and the previous chapter. The incentives are received by the network user in a number of ways:

- Price signals: the network user has to pay a higher price for power flows that do not support the grid. For example: a higher price for consumption when there is a shortage of production in the grid; in the same way, the network user will receive a lower price for production where there is a surplus of production. In extreme cases, the network user might even have to pay for production.

- Curtailment requests: the network user is asked to change power flow in a specific way: for example, to reduce production by 25%. There would typically be a financial incentive linked to that as well, but not necessarily.

- Curtailment orders: the network user is ordered to change the power flow in a specific way: for example, switch off all production on customer-side of the meter. This would typically be as part of an operating-reserve market in which the network user is voluntary participating. But although the participation is voluntary, the actual curtailment is compulsory for participants. The curtailment order may also be part of the conditions to connect to the grid.

Next to these incentives, there is also direct curtailment where equipment within the customer premises can be curtailed by one of the players on the electricity market, e.g., by a demand-response aggregator or by a distribution-network operator. Some of these direct curtailment signals can be blocked by the network users, others not. In the former case, this is, in fact, a curtailment request but with the default answer to accept the request. Direct curtailment that cannot be blocked is equivalent to a curtailment order.

5.7.2 SMALL CONSUMERS

Small consumers, domestic and small commercial, have access to a range of equipment that can be turned on or off in response to incentive signals or that can be part of a curtailment program. A subdivision partly based on, among others Brooks et al. [2010], is between four groups of consumption:

- Consumption whose absence is noticed by the customer immediately and where the impact continues after the consumption is restored. Desk-top computers are an example of this; they require restart and data may be lost when the computer shuts down. With many home appliances, like washing machines or dish-washers, a shut-down while they are being used will require a restart of the program. Also a shut-down of the oven will typically require that one has to start cooking again from the beginning. With some consumption, there is even a safety issue, like with lighting on a stair-case.

- Consumption whose absence is noticed by the customer immediately but where the impact is mainly limited to the period that the consumption is interrupted. Most lighting, television and radio, an external screen with a laptop, are examples of such consumption.

- Consumption whose absence is noticed by the customer only after a certain time. Heating and cooling are typical examples; more about this later.

- Consumption that can easily be shifted towards a later moment, even several hours away, without the customer being impacted. Charging of all kinds of batteries falls in this group, in the future especially, the batteries for electric vehicles,. But also the use of the washing machine and the dish washer can often be shifted to another time of day or even to another day. According to Brooks et al. [2010], up to 33% of loads has the possibility to be included in some kind of demand reduction without a significant impact on end users.

Heating and Cooling

Heating and cooling requires some further discussion. The main heating and cooling load consists of space heating and cooling (air conditioning). When turning off the heat, a building will start to cool down; the exact cooling-down process is rather difficult to describe and involves not only the temperature difference between inside and outside of the building, but also additional cooling like wind and humidity and additional heating like insolation. Also opening of the door (to enter or leave the building) has a big influence. A good approximation however is to describe it as an exponential decay with a certain time constant, the "thermal time constant" of the building. For example, when the thermal time constant is 25 hours, the difference between inside and outside temperature will drop by 4% (one 25th) every hour. The same holds for the heating up of the building in summer after turning off the air conditioning.

The thermal time constant of a building varies significantly, and it is rather difficult to get accurate information on this. The information available holds for well-insulated buildings and does typically not include influence from wind or insolation or opening of doors. The values cited for the thermal time constant vary between 25 and 60 hours.

Let us consider two extreme examples (very low temperature and very high temperature) to see what this would imply for the use of heating and cooling in demand-response schemes. The reason for using extreme temperatures is that these are the situations when the need for reducing consumption is most likely needed.

Assume an inside temperature of 23°C (73°F) and a very low outdoor temperature of −25°C (−13°F). For a four-hour interruption and a thermal time constant of 25 hours (the lower limit of the range for a well-insulated building), the temperature will drop by

$$\frac{4}{25} \times (23 - (-25)) = 8°C \tag{5.1}$$

After four hours, the temperature in the building would have dropped to 15°C (59°F). This is on the cold side but acceptable.

For a summer example, consider an inside temperature of 20°C (68°F, inside temperatures in hot climates tend to be lower than in cold climates, for some reason) and an outside temperature of 41°C (106°F). After a four-hour interruption, the temperature, for the same thermal time constant of 25 hours, has risen by 3°C to 23°. This would be completely acceptable.

But that was for well-insulated buildings, and there are still a lot of not-so-well-insulated buildings around. Assume instead a thermal time constant of 8 hours. For the winter example, the temperature will have dropped to just 4°C (39°F) after four hours. For the summer example, the temperature will have risen to 28°C (82°F). The winter case will be certainly unacceptable, but even the summer case might not be acceptable to the customer.

The conclusion is that a four-hour interruption of the heating or cooling is possible for a well-insulated building, but not for a badly-insulated building. Switching heating or cooling off for shorter periods is a solution acceptable for all individual customers, but the recovery peaks could make that the net reduction in consumption rather small. The net saving in energy consumption is mainly the lower heat loss with the environment due to the lower temperature difference. To be able to reduce the total consumption, there has to be a reduction in temperature difference, thus a change in indoor temperature.

There is also another look at this possibility. An incentive for turning off heating or cooling completely during several hours will be more attractive the better the insulation of the building. Customers with bad insulation will most likely not sign up to such a scheme with the result that their electricity costs will become higher. This, in turn, will provide an additional incentive for better insulation which will lead both to reduced peak consumption and reduced overall energy consumption.

Instead of turning off the heating or cooling completely, the thermostat setting can be impacted. This can again either be done manually or automatically. How this impacts the power consumption is shown in Figure 5.19. Immediately upon the change in setting (at time T_0), all heating and cooling will turn off and the consumption drops to zero. We assume that the change in setting is big enough; a change of less than a degree might not be enough. Also note that the figure only shows the consumption of the heating or cooling that is part of the curtailment scheme.

The consumption for an individual consumer will pick up again when the temperature has dropped below the thermostat setting. This will happen first for buildings with a low thermal time constant (at time T_1) and last for buildings with a high thermal time constant (at time T_2). After a while the consumption settles down at a lower value than before. The drop ΔP in consumption is proportional to the change in thermostat setting.

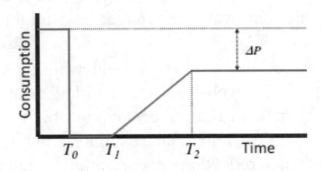

Figure 5.19: Drop in power consumption after a change in thermostat setting.

Consider a change of 4 degrees and the same temperatures and thermal time constants (8 and 25 hours) as in the earlier examples. For the winter case, we find T_1 = 40 min, T_2 = 2 hrs, and ΔP = 4%. For the summer case, the results are T_1 = 1.5 hrs, T_2 = 5 hrs, and ΔP = 19%. Especially for the summer case, the reduction is substantial even after several hours; for the first hour, the reduction is even 100%. Reduction of winter consumption is more difficult because the difference between indoor and outdoor temperature is more than twice is big.

Home Automation

For several of the schemes currently in place or part of demonstration projects, the customer receives plenty of notice about incentives. The hourly electricity price is available in the afternoon the day before. Also with most critical peak price and critical peak rebate schemes, notice is given the day before. Based on this information, the network user can plan the electricity consumption during the day, like shifting consumption to low-price hours whenever possible and switching off air conditioning or changing the thermostat setting.

But this manual intervention would require constant attention. An alternative is to obtain curtailment signals from the grid that can be blocked by the network users. But also this still requires some attention because curtailment is not always desired. A solution that could be adjusted to the wishes of every single customer is to integrate the control of electricity consumption in a kind of "home automation" together with a lot of other functions beyond the scope of this book.

When participating in balancing or real-time markets, or when offering operating reserve that has to be made available fast upon request, manual control is no longer possible. Some kind of control system, for example, as part of the home automation, would be needed.

A detailed description of home automation for domestic customers is presented by Lui [2010]: a central controller takes care of the different devices that contribute to the domestic consumption. Examples of appliances that can be controlled are dishwashers, cloth dryers and refrigerators. These appliances can be switched off without adverse impacting the customers for 60–90, 20–60, and 40–60 minutes, respectively. The biggest contribution to the peak consumption comes from the cloth dryer: above 3 kW; it will thus also give the biggest opportunity for reduction of the peak load.

However, the contribution of dishwashers, clothes dryers, and refrigerators to the total electricity demand at national level in the United States is only 0.11%, 0.08% and 0.05%, respectively. These maximum contributions do not even occur at the same time of day. When all such appliances would be part of a demand-response scheme, the total reduction in consumption for a larger area would still be only about 0.2%. The main gain of any demand response scheme will still be heating and cooling and in the future possibly the charging of electric vehicles.

5.7.3 SMALL AND LARGE PRODUCTION

When a network user has access to production, it is easier to react to price signals or curtailment requests or orders. But here also there are different types of production where it concerns their ability to react to incentives to reduce production.

- The consequence of reducing production is only the loss of income from the production, and increasing production is fast and easy. The decision in that case is easily made: when the price of electricity is too low (or too much negative in some cases), the production will be stopped. Most renewable electricity production falls in this category: sun, wind but also large hydro. Among the thermal units, only gas turbines belong here.

- The consequence of reducing production is the same as for the first category, but increasing reduction or restarting is a long and difficult process. This is the case for most large thermal production units.

- The reduction of production has other consequences beyond the loss of income. This is the case with most combined-heat-and-power units. The electricity production is linked to the heat demand; reducing electricity production will reduce the heat production. For domestic applications, the indoor temperature will drop (see the discussion earlier in this section); for industrial installations, the drop in heat production will typically cause a shut-down of the installation.

The above categorization holds for reduction of production. But in some cases, an increase in production is needed (in the existing grid this is actually more common). Only for some units is it possible to increase production on short notice. Hydropower installations and gas turbines are best equipped for this and are most often used for this. Wind-power and solar-power installations can operate with a certain reserve so as to increase

increase in the carbon-dioxide emission. Any market structure that would create incentives for such reserves would probably not be the most efficient one from an emission viewpoint.

Combined-heat-and-power(CHP) can in some cases increase production, for example, when it is combined with thermal storage. Also additional cooling will enable increased electricity production without increased heat production. This will result in reduced efficiency because the additional heat (40 to 60% of energy) will not be used. In fact, the marginal production will be rather equivalent to the production from a gas turbine. Allowing a CHP unit to increase production also requires additional investments in the unit. The participation of CHP units in the balancing market has been studied for Denmark and Germany; both countries have a large amount of CHP as well as wind power.

A customer may also decide to combine CHP with electric heating. When the electricity price is high the CHP unit is used. The income from selling the electricity will compensate the costs of the gas. When the electricity price is low, electric heating is used. This of course assumes that the gas price or the price of any other fuel used will not follow the same price patterns as the electricity price. The interaction between gas markets and electricity markets is a subject that has insufficiently been studied.

Another application of on-site production is the starting of back-up generation when the electricity price gets high. This is again a rather simple trade-off: when it becomes cheaper to produce electricity from the emergency generators than to obtain it from the grid, the emergency generators will be started. This will contribute to the price elasticity for industrial customers that have access to back-up generation. When the majority of those customers subscribe to hourly electricity prices, their price elasticity will make that the electricity price on the retail market will never get higher than the price of fuel for back-up generators.

5.7.4 STORAGE

When a network user has access to storage facilities on his side of the meter, the possibilities increase again much more. Storage can be used for both production and for consumption. The basic principle is rather simple: buy electricity when it is cheap and sell it when it is expensive. As long as the price differences are bigger than the losses in conversion and storage plus any loss of life of the installation due to cycling, a profit can be made. The details are somewhat more complicated and require careful planning to maximize the profit. The reason for this is that a storage installation is expensive and that the costs increase with storage size, sometimes even faster than linear. As a result, the owner wants to limit the storage capacity. Once the storage is filled with energy, no more cheap electricity can be bought, and once the storage is empty, no more expensive electricity can be sold. Selling and buying thus has to be based on the expected electricity prices. The spot-market prices are known ahead of time and can even be reasonably predicted a few days ahead. The price on the balancing market is very difficult to predict (if predictable at all), but it is the balancing market where differences between buying and selling are biggest, so that this is where the biggest profit can be made.

Storage is currently being studied in combination with solar and wind-power installations. Several applications for solar-power installations are presented by Hara et al. [2009]. In one demonstration project, 3 to 5-kW roof-top solar panels are installed with 550 domestic customers. A 9 kWh battery pack (4900 Ah) is installed with every customer to prevent overvoltages and grid overloading to the solar-power production. A 500-kW battery storage is installed together with a 2000-kW solar power installation, also in Japan. The storage here is used to compensate for intra-hour fluctuations in power production. Recently, a number of installations have been commissioned where battery storage is combined with wind-power installations. Battery storage is also studied as part of several microgrid projects and with electric vehicles. Both will be discussed below.

5.7.5 MEDIUM-SIZED AND LARGE CUSTOMERS

Industrial customers often have a number of different processes and devices that consume electricity. The costs of a complete non-planned interruption can be very high, but the costs of a planned interruption are typically much smaller. The price incentive at which demand response becomes attractive should be compared with the costs due to planned interruptions. These costs are on average a few USD/kWh with a large spread between individual customers. Current peak prices on the retail market are up to a few

USD/kWh. There is thus some potential for demand response of industrial customers during periods of high prices.

The published costs for planned interruptions consider the loss of supply to the whole installation. With demand response, the customer can pick parts of the installation that have less immediate impact while keeping the important processes running. Whenever there is storage of intermediate or end products somewhere, this offers a possibility for reducing consumption when electricity price is high. The options vary strongly between installations, but in all cases, participation in demand response schemes requires some organization. The additional costs are however small for large industrial customers. Automation and a control room are normal parts of such installations; the amount of investments needed is small.

With commercial customers, the air-conditioning (heating in winter) offers the main opportunity for demand response. The discussion above on thermal time constants also applies to commercial customers. Some commercial customers have large refrigeration installations. These offer some opportunities as well, but the application could be limited when, for example, temperate ranges are strictly controlled as part of food-and-health regulations.

5.7.6 ELECTRIC VEHICLES

The fast introduction of large numbers of plug-in electric vehicles would be a huge challenge to the grid. Without any coordination of their charging, overloads will occur especially in rural distribution networks. A transition from gasoline-driven to electrical vehicles will thus require either significant investments in the grid or a solution that reduces the impact of vehicle charging on the grid.

There are some issues with harmonic emission due to vehicle charging, but we will not address those here. Instead, we will only address solutions for limiting the contribution of vehicle charging to the peak load of the grid. Whenever the loading of the grid is less than the peak loading, any additional consumption will not be of much concern. The only impact is an increase in losses.

The daily energy use of an electric vehicle is estimated to be on average 10 kWh per day with charging typically requiring 2 to 5 hours every day, thus a consumption of 2 to 5 kW. Fast charging takes an even higher active power. The vehicle is however plugged in 10 to 15 hours per day at home and another 8 to 10 hours at the office [Brooks et al., 2010]. In Ungar and Fell [2010], the total consumption due to electric vehicles is estimated for 2017 in ten metropolitan areas. By spreading the charging over an eight hour period, the maximum consumption due to the electric vehicles is reduced to 25% of the non-controlled maximum. When spreading over 12 hours, it is even brought down to only about 15%.

A large amount of research is on-going for methods to reduce the contribution of electric cars to the peak load in the grid. These methods are broadly referred to as "smart charging". With some of the methods being discussed, the network operator can block the charging when the total consumption (charging plus everything else) gets too high. This solution is very similar to the use of curtailable electric boilers or air-conditioning units as are in place at several locations and under study at more sites. When a customer has subscribed to hourly electricity prices, critical peak prices, or critical peak rebate, there is a strong incentive to charge the vehicle outside of the peak hours (i.e., when electricity is cheaper). No additional incentive schemes are needed for this.

A specific line of on-going research concerns the control of numbers of electric cars. These cars could be all at the same location or at different locations. Examples of many cars at one location are, of course, parking places or garages that offer charging facilities, but also companies that offer these facilities to their employees, parking spaces with apartment buildings, or companies where the use of cars is their main activity (like taxi companies). But the cars could also be spread over many locations, like controlled charging spread over a whole city or over large parts of the countryside.

Several different constructions are being studied, but generally speaking, the owners of the vehicles have a contract with a "charging provider". This charging provider has the same role as the electricity retailer but only for electric vehicles. Each customer, upon connecting the car, indicates how much the car should be fully charged and before which moment in time (for example: fully charged tomorrow morning at 7, or charge at full charging capacity for 20 minutes). The charging provider schedules the charging such that the costs are minimized. In some cases, also, the prevention of grid overloads is considered in the scheduling of the charging. In the future, grid overloads can also be prevented through a market, for example, as discussed in Section 5.5.

The charging provider would base its main scheduling on the prices on the day ahead market. Next to that, the charging provider would likely participate in the balancing market so as to further reduce the costs. This assumes that it is possible to find a way of distinguishing between the consumption on the spot market and on the balancing market. This could be solved by making charging providers also balance responsible, which would require that they actually put buying bids on the day-ahead market and have to ensure that they follow the predicted consumption.

Other studies specifically consider charging providers being active on the balancing market. According to Brooks et al. [2010], all balancing could come from electric vehicles in just 20 years' time. For PJM, an independent system operator for a region with about 58 million customers, this would require 3.2 million vehicles. Similar studies done in Germany conclude that participation of 3 to 5% of all vehicles would be enough for the balancing market [Dietz et al., 2011].

Optimal participation of electric vehicles in balancing markets would require not only a scheduling of the charging but also occasional discharging of the batteries when the price for electricity is very high (high up-regulation price on the balancing market). This is often referred to as "vehicle to grid" or "V2G".

5.7.7 MICROGRIDS AND VIRTUAL POWER PLANTS

Two important smart-grid developments at the moment are "microgrids" and "virtual power plants". In both cases, a number of production units, possibly together with consumption and storage, act as one unit where the aim is to control the total production or consumption. The main difference is that a microgrid is connected to the grid at one location whereas a virtual power plant is connected at multiple locations. The terms, are, however, somewhat broadly used: e.g., the combination of a hydropower and a wind power installation with one connection point to the grid is also typically referred to as a virtual power plant. Even microgrids are sometimes referred to as virtual power plants.

A specific application of microgrids is their ability to operate independently from the grid. This application is the subject of the majority of the research on microgrids. It requires, however, significant additional investments in protection and control beyond those needed for participation in most electricity markets [Kroposki et al., 2008]. The main advantage of this is an increase in reliability for the network user. Island operation during interruption of the public supply has been used for many years by industrial installations as a way of increasing reliability. What is new with the developments on island operation of microgrids is the use of modern technology, including renewable energy resources, small-scale combined-heat-and-power, fuel cells, battery storage, curtailment of consumption and power electronics. Several demonstration projects are in place around the world.

The island-operation option is not a necessity for a microgrid to participate in the market. Adding this option will only be profitable at locations with a low reliability. In that case, controlled island operation will provide an additional advantage for installing local production. Especially, small-scale CHP has good possibilities for island operation. When the production is in the form of solar or wind power, it has to be combined with storage and/or curtailment of consumption to enable island operation.

The different options for controlling production and consumption are similar to the ones discussed in Section 5.7.2 and Section 5.7.3. However, a microgrid has a larger range of options for control so that the opportunities are bigger. An interesting example of an industrial microgrid is the one in Aichi, Japan. This installation contains seven fuel cells (total capacity 1400 kW), a sodium-sulfur battery (500 kW capacity) and 300 kW of solar panels. The consumption consists of several commercial buildings including an exhibition center. Several demonstration projects have also been started in Europe.

5.7.8 SMART METERS AND OTHER COMMUNICATION

The communication between a customer and the grid (the utility) used to be only through the electricity meter. This one was read regularly (or, occasionally), and the customer received the bill for the electricity consumption. Time-of-use tariffs caused some complication; this was solved by giving the customer two meters (or two counters within one meter) that were active during different times of the day. The change from one tariff to the other could either be based on a clock in the meter or, more commonly, by sending a signal to all meters via power-line communication. The meter reading, however, still took place manually. In fact, a large part of the electricity meters around the world still require manual reading.

Modern electronic meters ("smart meters") no longer require manual reading. Instead, the consumption is sent to the network operator (or to a separate metering operator) via a telecommunication link. Also here power-line communication is the most popular method. After all, whenever there is a meter, there is electricity. Some problems have been reported with power-line communication [Rönnberg et al., 2011], but overall the method works good enough for existing applications. When using more tight communication, like hourly meter reading or when many small customers start taking part in the balancing market, power-line communication may no longer be sufficient. Alternative ways of communication include radio links, telephone and internet.

Several of the curtailment programs involve sending information to the customer on prices. These prices and/or the hours for which they hold are in modern applications always published on the internet-page of the network operator or utility. The prices may also be available from a toll-free telephone number. With critical peak price and critical peak rebate schemes, the customer needs to be warned additionally. Here e-mail, text messages (SMS) and telephone are the channels commonly used. With some schemes, a display at home with the customer is used to inform when the scheme is active.

As seen in Figure 5.18, the amount of communication will increase a lot in the future. The metering will have to be much more detailed, going down to minutes when a customer takes part in balancing or real-time markets. Also price signals will continuously be updated, and new prices will have to be available almost continuously. Next to that, curtailment orders and requests as well as direct curtailment commands will arrive. Some of this information is very time critical, and it is important that a reliable communication channel is available at any time. Other information is less time critical and can be sent whenever a communication channel is available. The information from the meter to the metering operator is less time critical. It does not really matter if this information is sent a few hours later. Price signals are more time sensitive, with delays up to several minutes probably acceptable. When it concerns price information from the day-ahead market even an occasional one-hour delay would be acceptable.

When it comes to curtailment orders and requests as well as direct curtailment commands, these can be very time critical, and in some cases, a reaction within seconds is needed. This is not so much the case on the market for electrical energy, where one minute is about the lowest time scale of interest. But on some of the ancillary-service markets consumption and production have to be adjusted within seconds.

Durations of seconds or even minutes may not seem a big challenge to modern communication, and during most of the time, there are no serious problems expected due to communication delay. However, during emergency situations in the grid (the kind that happens only once in a few years), millions of signals may have to be sent at the same time. This is when a reliable communication system is essential. Most likely, the same kind of curtailment and demand-response methods will have to be applied to the communication system as to the electricity grid. When an emergency situation occurs in the grid, all available communication capacity will be allocated to the grid and any communication that is less important at that moment will be curtailed. Your telephone may not work for a few minutes, but the grid will survive.

The main discussion going on at the moment is between two fundamentally different ways of communication with the customer. One way is to feed all communication through the electricity meter. The meter will become a communication gateway between the customer and the various electricity markets. Several of the developments on smart metering, including at regulatory level, are going in this direction. The alternative is to use the meter strictly for metering of consumption and production, while all communication goes via home-area networks, local-area networks, internet, mobile telephone network, etc. Under this second alternative, each device on the customer premises could be linked to a home or local-area network and receive an IP address. Plans have also been ventilated to provide every device with a kind of telephone number and link them via the mobile telephone network. The home or building-automation system, but also a utility-microgrid controller or an ancillary-service provider, could communicate directly with the device whenever needed and permitted by the owner of the device.

CHAPTER 6

Discussion

In this chapter, three remaining issues will be discussed. The first one (in Section 6.1) concerns the way in which the network operation is paid for. There is especially concern among some stakeholders that the tariff mechanism will be a barrier against the introduction of the smart grid. We will not go into that discussion itself but will only present some relations between tariff regulation and investments decisions by network operators. Section 6.2 addresses the shift of reserves from the network to the consumers by comparing a number of investment alternatives in a network with a limited secure transport capacity to a small town. Although the example is given for a growth in consumption, the same alternatives and reasoning hold for introduction of new or additional production. Finally, Section 6.3 gives a somewhat personal view on the coming trends in distribution, subtransmission and transmission networks.

6.1 ECONOMICS OF NETWORK OPERATION

Network operation is a natural monopoly: the network user cannot choose the network operator, and there is no competition between network operators. In the modern structure of the electricity market, network operation is also a regulated monopoly: the regulator prevents that the network operator makes use of its natural monopoly. An example of regulation is the setting by the network operator of the network tariffs and other income.

6.1.1 TARIFF REGULATION

The income of the network operator consists of use of system tariffs (typically called "network tariffs") and connection fees. A connection fee is a one-off payment by a new network user to cover the costs made by the network operator to connect this user to the grid. This could include the connection of an electricity meter, building a new line, but also a general strengthening of the grid. The network tariff is paid by each network user typically monthly. Regulation might control the connection fee, the network tariff or the total annual income of the network operator. The way in which this regulation is implemented will determine the way in which an investment decision impacts the economy of a network operator.

Consider that a network operator can choose between existing technology and new technology for solving a certain problem in the grid. Both are expected to have the same advantages for the grid. The new technology is expected to be cheaper, but there are both technical and economic risks associated with the new technology. The costs for the new technology and its advantages are less certain than for the existing technology.

When the regulator sets the tariff (or a maximum value for the tariff), there is a clear incentive for the network operator to choose for the cheapest solution. The network operator would, with such a type of regulation, choose the cheapest solution despite the higher risk.

An alternative way of regulation is where the regulator approves the investments made by the network operator and calculates the tariff based on the investments and the operational costs. The total income would become the investments costs plus a reasonable return on investment and the operating costs plus a reasonable profit. Choosing the cheap solution would not improve the economy of the network operator; it might even reduce the profit and return on investment in absolute terms. Because of the higher risk with the new technology, the network operator will, in many cases, choose the cheap solution.

This assumes that the regulator will approve the investment. When the new technology is deemed sufficiently mature, the regulator may decide to only allow the network operator to recover the costs for the new (cheaper) technology. That will create an obvious incentive for the network operator to choose the new technology.

When the investment is needed because of the connection of a new network user, the situation is very similar. When the connection fee is only dependent on the actual costs of the investment, it is in the advantage of the network operator to choose the more expensive but low risk solution. When the connection fee is fixed or when there is a sufficiently-low maximum fee, it might be more advantageous to the network operator to choose the cheaper but higher-risk solution. This assumes that the network user, who stands for the costs, has no say in the matter. This only matters when the connection fee depends on the investment costs. In that case, the network user may be willing to accept a higher risk in return for a lower connection fee. The solutions based on intertrip and curtailment are typical examples of this.

6.1.2 PERFORMANCE INDICATORS

The regulator can create other incentives to impact investment decisions by the network operator. The additional regulatory instruments can be used to encourage the use of new technology where that offers a cost-effective alternative. The most direct influence is to make a certain investment compulsory. The introduction of smart meters is driven in that way in most countries; the network operators are obliged to install them. Alternatively, the regulator can set requirements on the network operator that can only be fulfilled by using new technology.

A common instrument used is "incentive-based regulation" where the income of the network operator depends on the value of certain performance indicators. The reliability indices are used for this in several countries: the better the reliability, the more the income of the network operator. This higher income will come from an increase of the network tariffs.

Similar types of incentive regulation are being discussed to encourage the introduction of new technology (smart grids). This has been one of the aims for developing the list of performance indicators presented in Section 2.6. Three of these indicators, related to the integration of new production into the grid, are discussed below.

The hosting capacity is the amount of electricity production that can be connected to the distribution network without endangering the voltage quality and reliability for other grid users. To calculate the hosting capacity, it is important that performance requirements for voltage quality and reliability are agreed upon.

The hosting capacity could also depend on the type of electricity production. This again makes that it is important to clearly define how the hosting capacity is calculated. Incorrect definition or calculation of the index could result in new technology increasing the actual hosting capacity but not the index.

When the hosting capacity indicator is used as a revenue driver, it should not create an incentive for excessive unnecessary investments in the grid. The indicator should give the right incentive towards the use of cost-effective technology.

Allowable Maximum Injection of Power without Congestion Risks in Transmission Networks

This index can be considered as a transmission-system equivalent of the hosting capacity. It can also be seen as the net transfer capacity from a (hypothetical) production unit to the rest of the grid. The condition "without congestion risks" should be interpreted as obeying the prescribed rules on operational security. This indicator can be calculated on an hourly basis, considering the actual availability of network components and the actual power flows through the network. This would result in an indicator whose value changes with time. The indicator can also be calculated as a fixed value under pre-defined worst-case power flows and a pre-defined outage level (e.g., N-1). The resulting value would give the largest size of production unit that can be connected without risking curtailment.

When using this indicator as a revenue driver, the same care should be taken as with the hosting capacity. The incentive mechanism should not result in excessive unnecessary investments and the method for calculating the index should not favor one technology above another.

Energy not Withdrawn from Renewable Sources due to Congestion and/or Security Risks

This indicator quantifies the ability of the network to host renewable electricity production. In that sense, it is similar to indicators like hosting capacity and allowable maximum injection of power. But whereas the latter two indicators only quantify the actual limits posed by the network, this indicator quantifies to which extent the limits are exceeded. The value of this index is determined afterwards, so that there are fewer approximations and assumptions needed than for the other two indicators. The calculation is rather similar to the calculation of energy not delivered, an indicator that is commonly used for continuity of supply. The main assumption to be made will be the energy that would have been produced during curtailment or disconnection of the production unit.

Another advantage of using the actual energy not withdrawn as an indicator, especially when used as a revenue driver, is that there is no risk of the network operator investing heavily in a network to be prepared for production capacity that never arrives. The associated disadvantage is that this indicator will give less incentive to invest before renewable electricity production is in place. This could result in the network being insufficiently prepared for a sudden increase in the amount of renewable electricity production.

6.2 RESERVE OPTIONS

Consider a simple example to see what the design options are both with new technology and with classical technology. Assume a small town that is supplied by two underground cables. The expected growth in consumption makes that the supply will no longer be secure during peak consumption within the near future. The two cables will remain sufficient to supply the whole consumption of the town, but when one cable is not available during the peak, the town cannot be supplied. The following options will be discussed below:

- Adding a third cable.

- Local energy storage or production owned by the network operator.

- Curtailment with compulsory participation.

- Curtailment with voluntary participation.

- Dynamic rating.

- Demand response by means of time varying pricing.

The main issue here is whether the reserves should be in the grid or with the network users. The issues that are discussed with this rather simple example appear in many of the design cases in the grid, from low-voltage all the way up to the highest transmission levels, for growth in consumption, for growth in renewable electricity production, and for large power transfers due to market opening. Planning of sufficient reserves to be able to cover the highest amount of consumption or production even when important components are not available is the main challenge of designing the grid.

A Third Cable

The classical solution is to build a third cable. The total transport capacity of any two of the three cables should be enough to supply the maximum consumption of the town. When the town is somewhat remote, this could actually be a rather expensive solution; in densely-populated parts of the country, obtaining the various permissions could be a lengthy process.

The risk with this solution is that the growth in consumption will be faster than the new cable can be built. That will only reveal itself if one of the other cables will be not available during a period of high consumption. Otherwise, the supply will not be secure, but there will not be any interruption. There is also an economic risk here: when the expected growth in consumption does not occur the investment would not have been needed.

Local Storage or Production

An alternative solution is for the distribution network operator to have storage or production available to cover the peak consumption. We refer here to production or storage owned and operated by the network operator, not production and storage owned by network users. Normally, i.e., when both cables are available, the local production or storage will be standing idle. Only when one of the cables is out of operation and at the same time the peak consumption occurs will the local production or storage take over part of the supply.

There are some regulatory issues with ownership of production units by a network operator, but we will assume here that those have been resolved. The argument in favor of ownership is that the production will only be used to prevent an interruption for the network users, not to be part in any of the open markets. When a market participant is willing to supply electricity with a local production unit, that production unit will get priority over the one owned by the network operator. A related solution is used by some network operators when there is only one cable or line to a group of customers, typically a small group of customers. Upon failure of this line or cable, or during maintenance, a generator truck will be used to supply those customers.

The costs of local production or storage could still be rather high, but it might be able to build them faster because the permission process would likely take less time. A number of generator trucks could offer a very flexible solution but somewhat less reliable than a production unit or storage installation on the spot.

Normally a cable would fail outside of the peak consumption period. In that case, a local generator can be started or a generator truck can be driven to the town. But when the failure takes place during peak consumption, the reserve has to be available fast. The time span is related to the thermal time constant of the cable. Assuming the overload protection will not trip the cable, the time to make the reserve capacity available would be several minutes or longer. Some kind of automatic start or remote start will most likely be needed.

Also here there is the economic risk that the investment is actually not needed. That risk would be less when a number of generator trucks are purchased. Those can be used wherever there is insufficient transport capacity to cover the peak consumption. The risk that the reserve is not available on time after failure of one of the cables will be bigger, however.

The main disadvantage with this solution is that the local production or storage will be used only a small percentage of the time and that it is not possible to make any estimation of when it will be used. Only when one of the two cables is out of operation and at the same time the peak consumption occurs is the local storage or production needed. This might translate into, on average, just a few hours per year. The same holds, in fact, for the building of the third cable; as long as local storage or production is cheaper than building a new cable, it is a cost-effective solution.

Curtailment with Compulsory Participation

Instead of investing in the grid (as with the two earlier solutions) some of the reserves could be shifted to the consumption. The economic disadvantage of the two earlier solutions (a third cable and local storage or production) becomes suddenly an advantage. The fewer hours per year that reduction of consumption will be needed, the less the impact on the network users will be. Here it should immediately be added, however, that the repair time of an underground cable can be rather long; several days or even weeks is not uncommon. Maximum consumption typically occurs during a few hours every day. The reduction in consumption will thus be needed for a few hours every day, several days or weeks in a row, and after that, there might not be any need for such a reduction for several years.

Upon failure of one of the cables, the network operator reduces consumption whenever needed to maintain the maximum consumption below the rating of one cable. This could be in the form of rotating interruptions, but using direct curtailment of certain equipment (like changing the thermostat setting of electric heating) significantly reduces the inconvenience for the customers.

The risk with such a scheme is clearly with the consumers, as the scheme is compulsory. But the scheme might not have to be activated very often, so that the risk is limited.

A scheme like this will require some kind of communication infrastructure to curtail the equipment when needed. Detecting the need for curtailment is relatively easy; all it requires is a current measurement and some decision logic similar to the one that is typically already in place for the overload protection. The main communication structure will be to get the curtailment orders to the end-user equipment. The future developments on smart meters and home automation are very important here. When a suitable communication infrastructure is in place, implementing a curtailment scheme could be easy and quick. That would provide a huge flexibility compared to having to build new cables or lines whenever there is a possible lack of sufficient reserve. But when the communication infrastructure has to be completely set up for every curtailment scheme, the costs are likely to be higher than the costs of a new cable.

A scheme like this will function only up to a certain amount of reduction in consumption. The inconvenience for the customers will become too big when the reduction gets too much. We will come back below with some further thoughts about that.

Curtailment with Voluntary Participation

Curtailment with voluntary participation combines market principles with control by the network operator over the reserves. The participation in the curtailment scheme is voluntary, but the reduction in consumption is compulsory once a customer has signed up. With a scheme like this, the network operator can make a good estimation of the available reserve at any moment in time. The main risk for the network operator is that the number of customers willing to sign up for a curtailment scheme is not enough for the required reserve. There will be a similar maximum-curtailment capacity as with the previous solution; above a certain level of curtailment, the inconvenience for the network users will become too large.

The investment costs for a scheme with voluntary participation are very similar to those with compulsory participation. The running costs of the voluntary variant will be higher for the network operator because the customers will have to be given some kind of incentive to sign up, for example, a reduction in the network tariff. Despite these higher costs, the voluntary scheme is preferable. The market mechanisms during the sign-up will result in only customers being curtailed for whom curtailment has a limited impact. Customers with high costs during interruptions or customers with safety issues will simply not sign up to the scheme.

Dynamic Rating

Dynamic rating of lines, cables or transformers is a way of allowing a higher transport capacity most of the time. The gain obtained from dynamic rating is much bigger with overhead lines than with underground cables, but there is some seasonal impact and even the humidity of the soil has some influence. It is however not certain that this higher transport capacity will actually be available when peak consumption coincides with one cable being out of operation. Therefore, a dynamic rating scheme will most likely be combined with a curtailment scheme. For the same amount of consumption, the curtailment will be activated much less often with dynamic rating than with fixed rating. The other way around, for a given maximum amount of curtailment (for example in hours per year), the maximum-permissible consumption is much higher for dynamic rating than for fixed rating.

Demand Response

This is a completely market-based solution where the consumer decides to reduce consumption or not based on the incentive given. It is up to the network operator to create a sufficient incentive for the network user to reduce consumption. Once a high consumption occurs while one of the two lines is out of operation, the network operator will either increase the network tariff ("critical peak price") or pay for reduction in consumption ("critical peak rebate"). The risk for the network operator is that the amount of reserve turns out to be insufficient. Especially when the overload situation occurs seldom, the actual reduction in consumption might be difficult to predict. The amount of reserve available depends on the price elasticity which is not well known and which will most likely vary often over time. The results presented in Section 5.3.7 indicate that the reduction in consumption is 10 to 30% without enabling technology like direct-curtailment or remote setting of thermostats, and up to 50% with enabling technology.

Hosting Capacity for New Consumption

To further illustrate the way in which growth in consumption can be accommodated by the network, consider a simple example. The existing peak consumption is 14MW, and each of the two cables can transport 15MW before getting overloaded. The hosting capacity for new consumption is 1MW (we neglect "details" like reactive power and losses.)

When introducing a simple curtailment or demand-response scheme, a reduction in peak consumption by 10% can be achieved. A maximum consumption of 15MW after reduction would correspond to 16.7MW before reduction, i.e., a hosting capacity equal to 2.7MW. A more extended scheme including enabling technology would result in 40% reduction, with a pre-reduction peak of 25MW being possible, a hosting capacity of 11MW. When the expected growth in consumption is more than 11MW, a third cable would be the preferred option; the hosting capacity is 16MW for that solution. For a slower growth in consumption, the extended curtailment or demand-response scheme will be in place already so that it can be used in combination with the third cable. The maximum peak consumption that this combination can cope with is 50MW.

6.3 INITIAL TRENDS

In the earlier chapters of this book, many new technologies and market schemes have been presented. In this section, a brief overview will be given of some of the initial trends that are expected or viewed necessary by the author. The trends will be presented in different parts of the power system starting at the lowest voltage levels, although there are several related trends.

6.3.1 DISTRIBUTION NETWORKS

The challenges at distribution level (the parts of the network that are operated radially or mainly radially, with voltage levels up to about 50kv) are the integration of new types of production and consumption. In some parts of the distribution network, reliability and voltage quality remain a challenge.

- Strengthening of the distribution network, more lines, cables and transformers will be unavoidable in some cases.

- Curtailment of production in areas with strong growth in production and curtailment of consumption in areas with strong growth in consumption. Curtailment should not be made compulsory but based on market principles, either with signing up or through demand response.

- Protection of medium-voltage networks will make increasing use of communication; the anti-islanding protection will be the first to make use of communication. Instead of using local measurement of voltage and frequency, the opening of any breaker in the distribution network will release an intertrip signal to all downstream production units. The communication infrastructure for curtailment can also be used for this.

- Storage at distribution level under the control of the network operator, to compensate fast fluctuations in production and consumption, as a flexible intermediate solution. Business models will appear that combine support for distribution-network operation with participation in wholesale and ancillary service markets.

- Voltage control in medium-voltage network will make more use of communication, and some of the local production units will contribute to the voltage control. Small production units connected to low voltage will trip or reduce production when they result in too high voltage magnitude for other end-user equipment.

- A discussion will start on the need for keeping strict limits on voltage magnitude and other voltage-quality parameters in parallel with a discussion on immunity and emission limits for end-user equipment. This will include the new types of disturbances introduced by new types of production and consumption.

- Requirements on small and medium-sized equipment will be restricted and be such that they can be taken care of by the equipment manufacturers.

- Improvements in reliability and voltage quality will continue to be made, with emphasis being on parts of the distribution network that have a low reliability. Solutions used will include underground cables, increased use of communication, and widespread monitoring with automatic analysis of the results.

6.3.2 SUBTRANSMISSION NETWORKS

With "subtransmission", we refer to the parts of the network at higher voltage levels that are operated meshed, but below the highest voltage levels covering the whole country. The voltage levels concerned are about 150kv. The main challenges here are the changing and largely unpredictable power flows that occur due to new sources of production and the opening up of the electricity market.

- The building of new lines or cables will continue, mainly to connect new wind parks. A growing part of the new connections will be underground cables. Harmonic resonances due to the cables will be damped by power-electronics control in the converters with the wind turbines.

- Dynamic line rating will be used to increase the transport capacity of the existing lines. This will be combined with curtailment schemes.

- The participation in curtailment schemes will be part of the negotiations about the connection fee for wind parks and for industrial installations connected to the subtransmission network. Existing installations will be given incentives to join such schemes.

- Power-electronics based shunt and series controllers will be used to control reactive-power flows and to more equally distribute the power flows over different paths. Some of the power-electronic solutions will be transportable; they can be moved to another location within a few months so as to offer a flexible solution.

- Overload protection will no longer trip the component that is overloaded because of the high risk of cascading outages. Instead, the cause of the overload will be removed by means of curtailment, either curtailment of production or curtailment of consumption.

- Operational reserves in the network will be reduced, and, instead, the network operator will rely on automatic curtailment as operating reserve.

- When the growth in production or consumption takes place at lower voltage levels, an infrastructure will be developed to communicate and distribute curtailment requests over many small network users.

6.3.3 TRANSMISSION NETWORK AND SYSTEM

This concerns the highest voltage levels and the largest production units. The main challenges here are the large-scale integration of renewable electricity production and the creation of open electricity markets covering large geographical areas.

- New transmission lines will still be built to remove bottlenecks in the existing network and to transport large amounts of power from renewable sources. Voltage levels above 400kv will appear or will grow to cover a large geographical area. Some ac lines will be converted to dc, especially where stability limits the transfer capacity.

- The transmission network will see an increasing use of HVDC lines and other power-electronics applications. The HVDC lines and networks that are appearing for submarine connections and between synchronous networks will be extended onshore.

- Dynamic line rating will be common practice for important lines where the thermal capacity sets the limit.

- A new look at power-system stability will appear. It will become easier to estimate stability margins and to detect the early stages of instability. A widespread use of monitoring devices, with phasor-measurement units playing a key role, combined with automatic data analysis, will form the base for this. The result will be that smaller reserve margins can be used with higher transport capacity as a result.

- A new look at operational reserves will appear; the amount of reserve will be dependent on system and weather circumstances. Stochastic operational-risk assessment will play an important role in deciding the amount of reserve needed. Part of the operating reserve will be shifted from the transmission network to the network users by means of large automatic curtailment schemes. Participation in these schemes will be voluntary and based on market principles.

- A large number of consumers will be exposed to hourly prices; the resulting demand response will reduce congestion in the transmission system. Initially, the uncertainty in price elasticity will result in a larger volume on the balancing market. The participation of small and mediumsized network users in the balancing market, through aggregators, will increase at the same time.

- Further ancillary-service markets will appear to support the transmission-system operator. These markets will involve large production units as well as small units and demand response or curtailment. Small production units and consumers will be involved through aggregators or retailers.

6.3.4 NETWORK USERS

Much of this will only be possible when the network users become actively involved either by means of curtailment or through various market mechanisms. Communication between the market and the network users is essential for this. The two trends mentioned in Section 5.7.8 are important in this context: enabling communication between the grid and the network user (e.g., through the meter) and enabling communication between the grid and end-user equipment (e.g., through the internet). Once such a communication infrastructure is present a lot of the mentioned curtailment and demand-response schemes can be introduced at relatively small extra cost. This will result in a very flexible power system that is able to cope with the challenges mentioned at the beginning of this book. Demand response and curtailment can be used to avoid having to build new primary infrastructure or as a temporary solution before new infrastructure is available.

The introduction of new markets allowing a more active participation of the network users will also result in development on customer-side of the meter: home automation and storage are two of the possible developments. The various wholesale markets (day-head, intraday, balancing, future ancillary-service markets) will see the appearance of new types of participants like aggregators, microgrids and virtual power plants.

CHAPTER 7

Conclusions

The previous four chapters presented a range of different technologies, methodologies and market structures to address the challenges faced by the electricity network. Some of these solutions are already commercially available but not widely used yet; others still require further development and even research.

In the transition from the existing grid to the future grid ("the smart grid"), there will be driving forces from two sides: from the side of the challenges and from the side of the solutions, where many solutions will introduce new challenges. It is important to link solutions and challenges before deciding on further research and development and on demonstration projects. It is also important to realize which challenges are specific for one particular location and which ones are general.

This book did not go into technical details of the various solutions presented, although the author is very much aware that this is where the main new development and most of the work will take place. This includes further developments of power-electronics control and operational methods for the power system with power-electronic devices, reliable and secure communication infrastructures and protocols, interoperability standards to allow all kinds of equipment to communicate with each other, market structures for electricity price and network tariffs and economic models to predict the market performance, monitoring technology and tools for automatic analysis of large amounts of measurement data, new operational rules for the transmission system including stochastic methods, new design rules for subtransmission and distribution networks including mathematical tools for quantifying the risk carried by the different network operators, storage technology plus its application in the grid and on customer side of the meter, new mechanisms including standards and regulation for maintaining acceptable voltage and current quality, microgrids and virtual power plants, and much much more.

With all these new developments and possibilities, it is important to always keep in mind the most important driving force: the transition to a sustainable energy system.

Bibliography

ABB, *Transformer Handbook*, 3rd Ed., ABB Management Services, Zürich, 2007. Cited on page(s)

D. Abbott, Keeping the energy debate clean: how do we supply the world's energy needs, *Proc. IEEE* 98(1):42-66 (January 2010). DOI: 10.1109/JPROC.2009.2035162 Cited on page(s) 7

E. Acha, V.G. Agelidis, O. Anaya-lara and T.J.E. Miller, *Power electronic control in electrical systems*, Newnes, Oxford, 2002. Cited on page(s) 39

T. Ackermann (editor), *Wind power in power systems*, Wiley, 2005. Cited on page(s) 11

AEMO, *Multiple generation disconnection and under frequency load shedding - Thursday 2nd July 2009*, Australian Energy Markets Operator, 2009, http://www.aemo.com.au/reports/0233-0001.pdf Cited on page(s) 64

I. Albizu, E. Fernández, A.J. Mazón, J. Bengoechea and E. Torres, "Hardware and software architecture for overhead line rating monitoring", *IEEE Trondheim Power Tech*, June 2011. Cited on page(s) 40

P.M. Anderson, *Power system protection*, IEEE Press, 1999. Cited on page(s) 65

J. Arrillaga, *High voltage direct current transmission*, 2nd Ed., The Institution of Electrical Engineers, London, 1998. Cited on page(s) 38

J. Arrillaga, Y.H. Liu and N.R. Watson, *Flexible power transmission - The HVDC options*, Wiley, 2007. Cited on page(s) 38

A. Badano, "Regelverk intelligenta elnät och smarta mätsystem - Omvärldsanalys" (Rules and regulation on intelligent networks and smart metering - status in other countries, in Swedish), Annex B with report EI R2010:18, Energimarknadsinspektionen, December 2010. http://www.ei.se Cited on page(s) 98, 99

A. Baggini (editor), *Handbook of power quality*, Wiley, 2008. Cited on page(s) 23

D.E. Bakken, A. Bose, C.H. Hauser, D.E. Whitehead, G.C. Zweigle, "Smart generation and transmission with coherent, real-time data", *Proc. IEEE* 99(6):928–951, June 2011. DOI: 10.1109/JPROC.2011.2116110 Cited on page(s) 65, 81

R.E. Barlow, *Engineering reliability*, Society for Industrial and Applied Mathematics, Philadelphia, PA, 1998. Cited on page(s) 41

K. Bhattacharya, M.H.J. Bollen and J.E. Daalder, *Operation of restructured power systems*, Kluwer, 2001. Cited on page(s) 91

R. Billinton and R.N. Allan, *Reliability evaluation of power systems*, 2nd Ed., Plenum Press, New York, 1996 Cited on page(s) 41

M.H.J. Bollen, *Understanding power quality - voltage sags and interruptions*, IEEE Press, New York, 2000. Cited on page(s) 23, 54

M.H.J. Bollen and I.H.Y. Gu, *Signal processing of power quality disturbances*, Wiley / IEEE-Press, 20066. Cited on page(s) 23, 80

M.H.J. Bollen, L. Wallin, T. Ohnstad and L. Bertling, "On Operational Risk Assessment in Transmission Systems - Weather Impact and Illustrative Example", *Int. Conf. Probabilistic Methods Applied to Power Systems, PMAPS 2008*, Puerto Rico. Cited on page(s) 43

M.H.J. Bollen and F. Hassan, *Integration of Distributed Generation in Power Systems*, Wiley - IEEE Press, 2011. Cited on page(s) 11, 25, 48, 60, 61, 68, 75, 76, 80, 81, 123

M.H.J. Bollen and N. Etherden, "Overload and overvolage in low-voltage and medium-voltage networks due to renewable energy - some illustrative studies", *Innovative Smart Grid Technologies Europe*, Manchester, UK, December 2011. Cited on page(s) 75

S. Braitwaith, "Behavior Modification", *IEEE Power and Energy Magazine* 8(3):36-45 (Maj/June 2010). DOI: 10.1109/MPE.2010.936348 Cited on page(s) 97, 101, 102

A. Brooks, E. Lu, D. Reicher, C. Spirakis and B. Weihl, "Demand Dispatch", *IEEE Power and Energy Magazine* 8(3):20-35 (Maj/June 2010). DOI: 10.1109/MPE.2010.936349 Cited on page(s) 126, 127, 132, 133

P. Caramia, G. Carpinelli, P. Verde, *Power quality indices in liberalized markets*, Wiley, 2009. Cited on page(s) 23

CEER, *4th benchmarking report on quality of electricity supply*, Council of European Energy Regulators, Brussels, 2008. http://www.energy-regulators.eu. Cited on page(s) 23, 26

CEER, *Status review of regulatory approaches to smart electricity grids - A CEER Status Review Paper*, Council of European Energy Regulators, Brussels, 2011. Ref: C11-EQS-45-04, http://www.energy-regulators.eu. Cited on page(s) 32

CENELEC, *Requirements for the connection of micro-generators in parallel with public low-voltage distribution networks*, EN 50438, December 2007. Cited on page(s) 60

CENELEC, *Voltage characteristics of electricity supplied by public electricity networks*, EN 50160, April 2010. Cited on page(s) 26

S. Chowdhury, S.P. Chowdhury and P. Crossley, *Microgrids and active distribution networks*, The Institution of Engineering and Technology, London, 2009. Cited on page(s) 52

CIGRE (2010), "Voltage dip immunity of equipment and installations", final report by CIGRE/CIRED/UIE joint working group C4.110, CIGRE Technical Brochure TB412. Cited on page(s) 26

D. Corbus, D. Lew, G. Jordan, W. Winters, F. Van Hull, J. Manobianco and B. Zavadil, Up with wind - Studying the integration and transmission of higher levels of wind power, *IEEE Power and Energy Magazine* 7(6):36–46 (November/December 2009). DOI: 10.1109/MPE.2009.934260 Cited on page(s) 36, 45

J. Deuse and G. Bourgain, *Results 2004–2009, Integrating distributed energy resources into todays eletrical system*, ExpandDER, 2009 http://www.eu-deep.com Cited on page(s) 52

B. Dietz, K.-H. Ahlbert, A. Schuller and C. Wienhardt, "Economic benchmark of charging strategies for battery electric vehicles", *IEEE Power Tech Conference*, Trondheim, June 2011. Cited on page(s) 133

J. Driesen, T. Green, T. Van Craenenbroeck and R. Belmans, "The development of power quality markets", *IEEE Power Engineering Society, Winter Meeting*, January 2002. DOI: 10.1109/PESW.2002.984996 Cited on page(s) 121

J. Driesen and F. Katiraei, "Design for distributed energy resources", *IEEE Power and Energy Magazine* 6(3):30–39 (May/June 2008). DOI: 10.1109/MPE.2008.918703 Cited on page(s) 52

R.C. Dugan, M.F. McGranaghan, S. Santoso, H.W. Beaty, *Electric power systems quality*, 2nd Ed., McGraw-Hill, 2003. Cited on page(s) 11, 23, 25

J. Endrenyi, *Reliability modelling in electric power systems*, Wiley, 1979. Cited on page(s) 41

ENTSO-E, *Draft Requirements for Grid Connection Applicable to all Generators*, ENTSO-E, 22 March 2011, http://www.entsoe.eu Cited on page(s) 61

ERGEG (2010a), *Position paper on smart grids - An ERGEG conclusions paper*, European Regulatory Group on Electricity and Gas (ERGEG), Ref: E10-EQS-38-05, 10 June 2010. http://www.energy-regulators.eu Cited on page(s) 30

ERGEG (2010b), *Pilot Framework Guidelines on Electricity Grid Connection*, European Regulatory Group on Electricity and Gas (ERGEG), Ref: E10-ENM-18-04, 7 December 2010. http://www.energy-regulators.eu Cited on page(s) 61

N. Etherden and M.H.J. Bollen, "Increasing the hosting capacity of distribution networks by curtailment of renewable energy resources", *IEEE Trondheim Power Tech*, June 2011. Cited on page(s) 40

B. Fox, D. Flynn, L. Bryans, N. Jenkins, D. Milborrow, M. O'Mallay, R. Watson and O. Anaya-Lara, *Wind power integration - connection and system operational aspects*, The Institution of Engineering and Technology, London, 2007. Cited on page(s) 11

Thomas L. Friedman, *Hot, Flat and Crowded - Why the world needs a green revolution and how we can renew our global future*, Chapter 12. The energy internet: when IT meets ET, Penguin Books, London, 2009. Cited on page(s) 7

H. Gharavi and R. Ghafurian, "Smart Grid: The Electric Energy System of the Future", *Proc. IEEE* 99(6):917–921, June 2011. DOI: 10.1109/JPROC.2011.2124210 Cited on page(s) 3

M. Goldberg, "Measure twice, cut once", *IEEE Power and Energy Magazine* 8(3):46-59 (Maj/June 2010). DOI: 10.1109/MPE.2010.936351 Cited on page(s) 100, 108

T. Gönen, *Electric power distribution system engineering*, McGraw-Hill, 1986. Cited on page(s) 33

T. Gönen, *Electric power transmission system engineering - Analysis and design*, Wiley, 1988. Cited on page(s) 33

K. Hamilton and N. Gulhar, Taking demand response to the next level, *IEEE Power and Energy Magazine* 8(3):60-66 (Maj/June 2010). DOI: 10.1109/MPE.2010.936352 Cited on page(s) 102, 103

R. Hara, H. Kita, T. Tanabe, H. Sigihara, A. Kuwayama, S. Miwa, Testing the technologies Demonstration grid-connected photovoltaic projects in Japan, *IEEE Power and Energy Magazine* 7(3):77–85 (May/June 2009). DOI: 10.1109/MPE.2009.932310 Cited on page(s) 131

N. Hatziargyriou, "Microgrids - the key to unlock distributed energy resources?", *IEEE Power and Energy Magazine* 6(3):26–29 (May/June 2008). DOI: 10.1109/MPE.2008.920383 Cited on page(s) 52

N.G. Hingorani and L. Gyugyi, *Understanding FACTS - Concepts and technology of flexible AC transmission systems*, IEEE Press, 2000. Cited on page(s) 39

S.H. Horowitz and A.G. Phadke, *Power system relaying*, Research Studies Press - Wiley, 1995. Cited on page(s) 65

IEC (2008), Electromagnetic Compatibility, Part 3, Section 6. Assessment of emission limits for distortion loads in MV and HV power systems, IEC 61000-3-6, February 2008. DOI: 10.1109/PESW.2000.847188 Cited on page(s) 35

IEC (2009), Electromagnetic Compatibility, Part 3, Section 2. Limits for harmonic current emissions (equipment input current ≤16A per phase), IEC 61000-3-2, April 2009. DOI: 10.1109/PESW.2000.847188 Cited on page(s) 35

IEEE (1993), *IEEE Guide for liquid-immersed transformer through-fault-current duration*, IEEE Std. C57.109-1993. Cited on page(s) 73

IEEE (2003a), *IEEE Guide for Electric Power Distribution Reliability Indices*. IEEE Std.1366-2003. Cited on page(s) 26

IEEE (2003b), *Standard for Interconnecting Distributed Resources with Electric Power Systems*. IEEE Std. 1547-2003. Cited on page(s) 60

M.Z. Jacobson and M.A. Delucchi, A path to sustainable energy by 2030, *Scientific American* 301(5):38-45 (November 2009). Cited on page(s) 7

N. Jenkins, R. Allan, P. Crossley, D. Kirschen and G. Strbac, *Embedded generation*, The Institution of Electrical Engineers, London, 2000. Cited on page(s) 11

G. Johnson, Plugging into the sun, *National Geographic* 216(3):28-53 (September 2009). Cited on page(s) 7

Michio Kaku, *Physics of the future*, Chapter 5. Future of Energy, Pinguin Books, London, 2011. Cited on page(s) 7

A.K. Kazerooni, J. Mulate, M. Perry, S. Venkatesan, and D. Morrice, "Dynamic thermal rating applications to facilitate wind energy integration", *IEEE Trondheim Power Tech*, June 2011. Cited on page(s) 40

S. Kiliccote, M.A. Piette and J.H. Dudley, "Northwest open automated demand response technology demonstration project", Ernest Orlando Lawrance Berkeley National Laboraort, Environmental Energy Technologies Division, April 2009. Cited on page(s) 108

E.W. Kimbark, *Direct current transmission*, Wiley-Interscience, 1971. Cited on page(s) 38

B. Kroposki, R. Lasseter, T. Ise, S. Morozumi, S. Papathanassiou and N. Hatziargyriou, "Making microgrids work", *IEEE Power and Energy Magazine* 6(3):41–53 (May/June 2008) DOI: 10.1109/MPE.2008.918718 Cited on page(s) 52, 134

P. Kundur, *Power system control and stability*, McGraw-Hill, 1994 Cited on page(s) 41

E.O.A. Larsson, M.H.J. Bollen, M.G. Wahlberg, C.M. Lundmark and S.K. Rönnberg, "Measurements of high-frequency (2-150 kHz) distortion in low-voltage networks", *IEEE Transactions on Power Delivery* 25(3):1749–1757 (July 2010). DOI: 10.1109/TPWRD.2010.2041371 Cited on page(s) 27

T.J. Lui, W. Stirling and H.O. Marcy, Get smart, *IEEE Power and Energy Magazine* 8(3):66-78(Maj/June 2010). DOI: 10.1109/MPE.2010.936353 Cited on page(s) 129

P. Lund, "Cell controller pilot project - Intelligent mobilization of distributed power generation", *3rd Int Conf on Integration of Renewable and Distributed Energy Resources*, Nice, France, December 2008. Cited on page(s) 53

D.J.C. MacKay, *Sustainable energy - without the hot air*, UIT Cambridge Limited, Cambridge, UK, 2009. Cited on page(s) 7

N. Martensen, P. Lund and N. Mathew, "The cell controller pilot project: from surviving system black-out to market support", *21st Int Conf on Electricity Distribution (CIRED)*, Frankfurt, June 2011. Cited on page(s) 53

P. Massee and H. Rijanto, The optimum adjustment of motor protection relays in an industrial complex, *Microelectronics and Reliability* 35(9-10):1245–1256 (September/October 1995). DOI: 10.1016/0026-2714(95)99375-S Cited on page(s) 82

K.R. Padiyar, *HVDC power transmission systems*, New Academic Science, Turnbridge Wells, Kent, UK, 2011. Cited on page(s) 38

PowerCentsDC Program - Final report, September 2010. http://www.powercentsdc.org Cited on page(s) 98, 99, 100, 102

R. Pratt, Scalable demand response networks: results and implications of the Olympic Peninsula GridWise demonstration, *3rd Int Conf on Integration of Renewable and Distributed Energy Resources*, December 2008, Nice, France. Cited on page(s) 116

S.K. Rönnberg, M.H.J. Bollen and M. Wahlberg, "Interaction between narrowband power-line communication and end-user equipment", *IEEE Trans. on Power Delivery* 26(3):2034–2039 (July 2011). DOI: 10.1109/TPWRD.2011.2130543 Cited on page(s) 135

H. Sæle and O.S. Grande, Demand response from household customers: experiences from a pilot study in Norway, *IEEE T. Smart Grids* 2(1):90–97 (March 2011) DOI: 10.1109/TSG.2010.2104165 Cited on page(s) 108

J. Schlabbach, D. Blume and T. Stephanblome, *Voltage quality in electrical power systems*, The Institution of Electrical Engineers, London, 2001. Cited on page(s) 23

J.C. Smith and B. Parsons, "What does 20 percent look like? - Developments om wind technology and systems", *IEEE Power and Energy Magazine* 5(6):22–33 (November/December 2007). DOI: 10.1109/MPE.2007.906565 Cited on page(s) 44

J. Snowdon, R. Ambrosio and T.J. Watson, The Olympic Peninsula project, presentation at *Smart Grid for Smart Cities*, 3 February 2010. Cited on page(s) 116

Y.H. Song and A.T. Jones, *Flexible ac transmission systems (FACTS)*, The Institution of Electrical Engineers, London, 1999. Cited on page(s) 39

J.S. Thorp, M. Adamiak, H.N. Banerjee, J.A. Bright, T.W. Cease, D.M. Clark, E.M. Gulachenski, S.H. Horowitz, W.C. Kotheimer, L.L. Mankoff, A. Munandar, S.L. Nilsson, A.G. Phadke, R. Ramaswami, G.D. Rockefeller, R. Ryan, M.S. Sachdev, H.S. Smith, E.A. Udren and C.L. Wagner, "Feasibility of adaptive protection and control", *IEEE Transactions on Power Delivery* 8(3):975–983 (July 1993). DOI: 10.1109/61.252625 Cited on page(s) 48

UCTE, *Final report system disturbance on 4 November 2006*, Technical report, UCTE, 2006. http://www.ucte.org/ Cited on page(s) 64

E. Ungar and K. Fell, Plug in turn on and load up, *IEEE Power and Energy Magazine* 8(3):30-35 (Maj/June 2010). DOI: 10.1109/MPE.2010.936354 Cited on page(s) 20,

K. Verhaegen, L. Meeus and R. Belmans, "Development of balancing in the internal electricity market in Europe", *European Wind Energy Conference*, Athens, Greece, February 2006. Cited on page(s) 108

R. Walawalkar, via PowerGlobe, 31 May 2011. Cited on page(s) 46

I. Wangensteen, *Power system economics - the Nordic electricity market*, Tapir academic press, Trondheim, 2007. Cited on page(s) 91

K. Yang, M.H.J. Bollen and M. Wahlberg, "A comparison study of harmonic emission measurements in four windparks", *IEEE Power Engineering Society General Meeting*, Detroit, July 2011. Cited on page(s) 27

Xiaodong Yang, Gengyin Li and Ming Zhou, Optimal allocation of electromagnetic pollution emission rights in power quality markets, *Int Conf Power System Technology (PowerCon)*, 2006. DOI: 10.1109/ICPST.2006.321879 Cited on page(s) 121

Author's Biography

MATH BOLLEN

Math Bollen received the MSc and PhD degrees from Eindhoven University of Technology, Eindhoven, The Netherlands, in 1985 and 1989, respectively. He has among others been a lecturer at the University of Manchester Institute of Science and Technology (UMIST), Manchester, U.K., and professor in electric power systems at Chalmers University of Technology, Gothenburg, Sweden.

Currently, he is professor in electric power engineering at Luleå University of Technology, Skellefteå, Sweden, senior specialist at STRI AB, Gothenburg, Sweden, and technical expert at the Energy Markets Inspectorate, Eskilstuna, Sweden.

Math Bollen is one of the leading researchers in power systems, having published over 300 technical papers in journals and at conferences on a range of subjects. He has defined voltage dips as a research area and more recently harmonic distortion in the frequency range 2 to 150 kHz. He has also developed the "hosting capacity" concept as an important measure for the ability of the grid to accept new production or consumption. Math Bollen has published two textbooks on power quality, "understanding power quality problems" and "signal processing of power quality disturbances" and recently a third textbook on "Integration of distributed generation in the power system".

Math Bollen is a Fellow of the IEEE and recipient of the CIGRE Technical Committee Award.

Index

Printed in the United States
by Baker & Taylor Publisher Services